THE
FLOWERING
WILDERNESS

THE WORLD OF NATURE

THE FLOWERING WILDERNESS

UBERTO TOSCO

With a foreword by
Sir George Taylor DSc, FRS, FRSE, FLS, VMH
Director of the Stanley Smith Horticultural Trust;
formerly Director of the Royal Botanic Gardens, Kew

ORBIS PUBLISHING·LONDON

The photographs in this volume have been supplied by
Afrique Photo, M Andi, IGDA Archives, C Bevilacqua,
N Cirani, A Filippini, R Grassetti, A Margiocco,
Marka, G Mazza, C e Östman, P2, L Pellegrini,
P Popper, S Prato, F Quilici, SEF

Colour and form are the chief elements of this book, its first-class photographs bringing to life a spectacular variety of flowering plants. The giant cacti growing in arid lands yet bearing beautiful blooms, the magnificent colouring of lush tropical vegetation, the vivid flowers of Alpine mountains, the unimaginable insect-eating plants: all these and more are here in their grace and perfect detail. Some flowers, such as the opulent orchid, are already well known, even to people who live in temperate climates, but others are still strangers.

Yet both the familiar and the exotic hold a fascination for botanist and layman alike. Their beauty is like that of the ballet, but the blooms achieve a subtle, colourful rhythm denied to even the most graceful of humans. There is something primeval about the trunks and roots of great trees, with their rugged bark evoking the armour of prehistoric animals.

It is not popularly realized, either, that the colour, shape and fragrance of a flower represent, in the strictest sense of the word, a form of communication. They convey to insects an invitation; to animals a warning, so that they know instinctively by certain colours and perfumes that a particular species of plant is forbidden to them. This is one of the many ways in which survival of plant species has been ensured over countless years.

The message with which we are greeted in our excursion into 'The Flowering Wilderness' is a most enriching one, and I am pleased to be the starting point of this exotic and exciting journey.

Sir George Taylor DSc, FRS, FLS, VMH
Director of the Stanley Smith Horticultural Trust;
former Director, The Royal Botanic Gardens, Kew

Contents

List of plates

A world of beauty

The plants that live all around us are essential to our existence on this planet. Without them, animals and men could not survive, for quite simple ecological reasons, and without them certainly the world would not be so delightful a place as it is. Plants are beautiful and continually surprising in their variation of size, form, and colour, giving to our landscapes vast areas of green, and to our gardens and forests all the colours of the rainbow. We imagine the humid jungles of the tropics and the water-starved desert lands populated with exotic, colourful flowers, and bizarre, shapeless trees, and to some extent this vision is real; for among the plants that grow in foreign countries – by definition 'exotic' – are many that fit these descriptions, as the illustrations in this book will show.

But beauty in vegetation is not confined to exotic plants. In the USA the variation in climate – virtually the whole gamut from arctic to tropical – can support a corresponding variety of plant species and indeed, almost every country of the world has examples of strange and beautiful flowers, often unknown to most people, who have little idea where to find them. It is true that temperate vegetation is less spectacular than that in other regions, although this may be partly the reaction of those who are too familiar with it, and less space will be devoted to it here than to the colourful and the unusual. However, the object is not merely to pick out a few outstanding species at random. In each of the chapters of this book we have attempted to give the descriptions of the habitats and structure that are necessary for a full appreciation of the beauty of these plants.

Strange shapes and colours are typical of exotic tropical flowers, as can be seen in those of Ananas comosus, from the New World. Cultivated varieties of this plant produce the fleshy fruit that we know as pineapple

Classification and climate

The world's vegetation can be divided into three main types, corresponding to different climatic areas: *desert* plants, which have to survive extremes of either heat or cold, as well as drought; *forest* plants, which flourish under heat and rain; and finally, flora of the *grasslands* of the world, in which the climate (not easily classified) supports a variety of less highly specialized types. This division is obviously a very rough one, since there is an almost indefinite number of plants, and the types merge into each other, but it is useful to relate man to these three principal categories.

The desert environment is by definition hostile to human and animal life, simply because very little vegetation can flourish in it. A detailed study of tundra and hot desert vegetation shows that some plants can exist there, but they are of interest mainly for their unusual features, rather than because they are of any practical use. Forest, in the broadest sense of the word, is land covered by trees, and although useful to man it is almost always destroyed in the advance of human progress. Grasslands tend to be the most suitable for human habitation, because the natural state of the plant life there is more or less ideal from man's point of view.

In a continent the size of North America, covering as it does a wide range of latitude and longitude, there are examples of all these types of environment, and correspondingly a good representation of plant species. Even Britain, which is about 40 times smaller than the USA and has much less difference in climatic extremes, soil, and general topography, nevertheless has greatly varied vegetation. Perhaps it would be helpful to go into the variations in these groups of flora in rather more detail.

In general, tropical plants are better able to adapt themselves to more temperate climates rather than the other way round. This can be explained by the fact that in Tertiary times middle latitudes as we know them today were subjected to much higher temperatures. Conifers, which are typical of temperate climates, originated in a tropical one.

There are other important climatic differences which may be considered. Even where there is a heavy mean annual rainfall, this probably comes in a series of heavy storms, with long hours of sunshine and intense insolation. Another point to remember is that high winds are short-lived and not as strong as in middle latitudes, and as a result vegetation does not have to withstand

The leaves of tropical plants, as well as the flowers, are often high coloured, bearing strange and beautiful markings. This picture shows the leaves of the Caladium picturatum, an aroid of the Ceylon forests

excessive transpiration (loss of moisture).

The majority of the plants found in extreme conditions are evergreen, casting their old leaves and producing new leaves continually, these leaves being thickly cutinized, and so able to continue with the process of photosynthesis in spite of the intense insolation. Many of the leaves have long pointed tips – 'drip tips' they are called – and help in the dispersal of excessive water during the heavy rainstorms.

Many of the plants in this area, too, have developed abnormal growth forms – lianas with their twining habits, buttresses or stilt roots, and aerial roots. Some plants are epiphytes which live out their life attached to other plants. Parasitism is also common among plants, and some examples of these adaptations are given in greater

detail in some of the later chapters.

Perhaps the most interesting forms of plant adaptation are found in the life cycles of the so-called insectivorous plants which, although often capable of photosynthesis, can also obtain food from insects which they capture through their special mechanism.

It may well be that these various forms of plant adaptations also play a part in the phenomenon of endemism – the fact that certain species are found in localized and well-defined areas, often due to natural barriers preventing the spread of these particular species. Some are endemic to the isolated islands of New Zealand where the sea acts as a natural barrier and prevents an exchange of plants between New Zealand and the rest of Australasia.

3

Habitat and plant life

Every form of life needs certain vital environmental conditions. Apart from a few very rare exceptions, vegetation cannot survive in an environment without air to provide oxygen, water (or at least a minimal percentage of humidity), without at least a minimum amount of light; or in extremes of temperature, either too low or too high.

The vast majority of the plant world, however, needs a habitat that provides an abundance of at least two of the four main environmental factors: water, air, light, and heat. Even in the most arid and waterless desert, prickly bushes and succulent plants do manage to flower, as long as there is enough air, light, and heat to support vegetable life. In the darkness of caves and in undergrowth, where there is low light intensity, the rocks and the bark of trees are covered by moss, ferns and lichens, living on humus rich in nutrients and water. Certain plants thrive at high altitudes, such as the *Eritrichium nanum*, a tiny forget-me-not that colours the slopes of mountains: this type of plant is bathed in sunlight and obtains moisture from the snow.

Plants tend to live in highly organized communities which have developed over many

Left: A group of coconut palms on Ambalangodo beach, Ceylon. Such scenery is typical of tropical regions.
Below: Ravenala madagascariensis, with elegantly fanned leaves whose bases are filled with water. They are often plucked to provide drinking water, and as a result the plant has come to be known as the 'traveller's' or 'pilgrim's' tree

millions of years and given the Earth's vegetation its present-day appearance. Survival and reproduction depend on well-developed co-existence between different types of vegetation, and on the environmental conditions, to which adjustments have to be made. The diverse plant formations of frozen, rocky, and sandy deserts, arid steppes, tundras, Alpine pastures, forests, and savannas, are all due to the gradual and very slow process of natural selection. Clues to plant development are found in the fossils that testify to the existence of many species in remote geological ages. From their evidence, it is believed that the thick ice-cap of Greenland conceals the remains of quasi-tropical forests that used to grow there before the Equator and the Poles were in their present positions.

The varying nature of exotic as well as temperate vegetation is explained by a number of things: the most important of these is climate, but there are a great many other factors which determine the development of plant types. Among the climatic conditions are the varying degrees of insolation, wind speed (and hence soil erosion), rainfall, and temperature. All these factors have their effect on the soil, notably from the point of view of drainage, which is highly important in the development of plant life. Vegetation is also influenced by physical forces. Extreme climatic changes, volcanic eruptions, earthquakes, and floods may also produce drastic and wholesale changes in vegetation.

Although all trees in tropical rain forests, irrespective of locality, are basically of one morphological type, namely broad-leaved evergreen trees, the actual species differ according to the region. The differences to be found in the Congo Basin and the Amazon, for instance, are due to the separate evolution of plant life over a very protracted length of time.

Man is another cause of interruption, and from the earliest times has cut down forests to estab-

Tropical environments are often characterized by particular groups of vegetation. This scrub landscape, in the shadow of Mount Kilimanjaro, is largely composed of acacia, thorn and Gramineae (grasses)

lish grazing land, or has taken over the land for his own cultivated crops. Finally, the plants themselves are in strong competition. In their struggle for existence against one another they become eliminated in a sort of 'survival of the fittest' evolution: for instance, tall shady trees compete with small light-loving plants, which are eventually shaded out and destroyed.

All these different factors explain the varying nature of exotic and temperate flora, and the gradual movement of flora from one zone to another: from the Equator to the high mountains of Kenya or to the Sahara, from the Sahara to the Nile basin and Arabia, from the coast of Tripoli to Sicily, from the Algerian shores of Spain, and from Arizona to the Carolinas.

Plants in cold climates

Of all the environmental conditions that influence the growth and development of life on Earth, it is arguable that none is more important than temperature. Plants, unlike the higher animals, have no means of controlling their own internal temperature, and therefore their essential biochemical mechanisms are at the mercy of the climate. For this reason, an examination of the types of vegetation peculiar to different temperature environments reveals a considerable diversity, testifying to the ability of the basic plant structure to adapt to the prevailing conditions.

It scarcely need be said that hot climates are more hospitable to plants than cold climates are; but both can support plant life, provided that it is sufficiently well adapted. In very cold conditions, the problem is not, however, simply one of low temperature. Because the conditions are so bleak, with few other species able to survive, the ground is inevitably poor in plant nutrients; and even the water, essential to a plant's existence, is frozen. The wind blows harshly, the hours of dark and light are irregular, and there is seldom insect means of transport for sexual reproduction by pollination.

Of course the vegetation is poor; of course it is sparse. But to the occasional traveller – or even the resident – in the bleakest cold outposts of the world, the fact that any plant is able to survive is a comfort, and the splashes of green or colour stand out brightly in the ice and snow, surprising and delighting the eye.

In the very far north, where the temperature is always very cold and the summer is very brief – no more than a short interval breaking up the long Arctic night – the vegetation provides one of the most significant examples of poverty caused by severe ecological conditions.

Arctic vegetation is divided into two main groups: that of the south, or subarctic, and the north arctic. These two regions are separated by the tree line, the northernmost point at which forests grow. Even in the subarctic what sparse vegetation there is has had to adapt itself to the very harsh environment: low temperatures, virtually continuous daylight in summer, and darkness in winter, poor soil and frozen ground. The 'permafrost' is the characteristic feature of these regions: as little as four inches below the surface one finds that the ground is of a permanent brick-like consistency, and completely ice-bound. The life-cycle of the few flowering species is very brief: barely six weeks, starting in the middle of July, and ending in August. Flowers bloom extremely rapidly, and die down almost as speedily as they spring up. A few Orchidaceae, Ericaceae, Asteraceae, Ranunculaceae, Scrophulariaceae – mostly perennials which reproduce asexually – make a brief but welcome change from the otherwise uniform drabness of the Polar regions. Some plant life, however, is always present, in spite of the severity of the conditions. Mosses and lichens have managed to establish themselves more or less successfully in this domain.

Above the tree line the territory is even more barren: sedges and grasses are the dominant vegetation, with an underlayer of lichens and mosses. A few dwarf trees struggle for existence, such as the stunted willows and the dwarf birches that are found in Alaska.

The vegetation of cold climates is obviously not as fascinating nor as showy as that of warmer areas, but the fact that there is any vegetation at all is very remarkable. In order to survive against the onslaught of the dry icy winds and blizzards, most plants tend to group themselves into cushion-like formation.

Tundra plants have a special sort of beauty all of their own, which is best appreciated with the aid of a magnifying glass. The rose-coloured or brown spore-bearing capsules of mosses and sphagnums, and the red or brown spore-bearing

bodies of lichens, hanging like sparkling drops round each leafy thallus, then become really clear.

Little breaks the monotony of the tundra landscape, except for a few scattered mountains. The scenery is characterized by muted but beautiful colours, such as pink, lilac, rosy red, white and pale gold, during the brief summer.

The Antarctic is even colder than the Arctic, particularly in summer. Its vegetable life is poorer, and the species that inhabit this vast continent of ice and snow are far removed from those of the Arctic or Alps. Only two flowering plants exist, and these only in the Graham Land Peninsula: *Deschampsia antarctica,* and *Coloban-thus crassifolius.* In a few select sheltered places, where the snow melts in the summer, brightly coloured mosses and lichens can be found in sparse areas. Because of the exceptionally strong winds, the whole of the Antarctic is very deficient in woody plants.

11

Vegetation changes as the polar regions are left behind. In temperate climates, tundra gives way to what is called 'woodland tundra'. This consists of a few bushes with twisted branches mingling with taller grasses. In the north these are mainly pines and birch trees, bushy willows, and species of the bilberry family. These woodlands are a prelude to the true forests which, in the northern hemisphere, are made up of conifers – the forests of firs, larches, and pines which stretch from Canada over the whole of subarctic Eurasia.

Pillar-shaped cacti and the so-called 'candelabra' spurges are typical of arid zones. The plants, which are rich in water, often grow to a considerable size

The origin of tropical flowers

While survival of plants at the lowest end of the world's temperature spectrum is a tribute to their adaptability and hardiness, the life of those at the other end – in the hottest climatic regions – is one of luxury and brilliance. In the main, a hot climate is ideal for plant growth, particularly when one is talking of a humid tropical environment with large amounts of organic matter in the ground.

A high temperature presents a rather greater problem to plant life when the region is dry, however, as the sparsity of vegetation in deserts shows. It is necessary, therefore, to consider hot environments according to the other conditions that prevail, distinguishing forests, marshes, mountains, deserts, and so on. For that reason, before considering the plants of the tropics in detail, and the environment they inhabit, we will first examine their geographical evolution.

The study of past and present-day plants is divided into three categories: plant palaeontology, the science of past organic life based on fossils, plant geography and plant genetics. Any advanced research must take into account the evolution of plants from as far back as the Carboniferous period, approximately 250 million years ago. Prior to this time, very little life was to be found on land: what living organisms there were existed in the sea, and a long period of adaptation was necessary to enable them to become acclimatized to their new habitat. Like all plants, modern tropical vegetations originated so long ago that they are currently linked with parts of the world that do not correspond, either climatically or geographically, with the areas that they occupy today. Numerous theories have been formulated to explain the changing aspects of the Earth's vegetation since its very earliest origins, and the most realistic of these theories, despite widespread criticism of it, is perhaps the one formulated by Wegener in 1915. His hypothesis was that in the earliest geological eras, the Earth was

formed by a single continent, known as 'Pangea' (the Greek for 'single land'), which continued as a sole or almost complete entity up to the Carboniferous age.

In these primeval conditions, the Atlantic either did not exist or existed only as a small inland sea, and the shores of Africa touched those of America, in particular South America, while the Antarctic and Australia formed a block connected both to the southern tips of America and Africa and to the lower point of the great Indian peninsula. Thus, Africa – which was still one with the island of Madagascar – was connected at certain points with the western coasts of India. Over the millennia, however, the land masses separated and little by little the continents acquired their present-day configuration and relative positions.

What is known as the 'migration of the Poles' took place in the same period, over thousands of years. The North Pole, which used to be in the middle of an immense ocean surrounding the western side of Pangea during the Carboniferous period, moved to what is now Alaska and from there reached the position in which it is today. The South Pole, on the other hand, occupied a position corresponding to present-day South East Africa before it migrated into the middle of the Antarctic.

Naturally, the Equator moved roughly in the same way as the Poles. During the Carboniferous period, it crossed almost the whole of central Europe, which was at that time covered with forests of sigillaria, lepidodendron and calamite, all plants that have now disappeared leaving only fossils behind; some notable examples are to be found in the fossil coal strata in the Ruhr.

From this brief history, it can be seen that a tropical flora already existed 250 million years ago – although its species were very different from those of today – and that it followed the inter-tropical, equatorial belt in its movements,

leaving in its wake fossils by which its existence can be identified.

As the Equator gradually moved southwards across the Mediterranean and took up its present position, the flora of the tropics changed slowly but continually, losing some of its characteristics and acquiring others, either through evolutionary processes or through adaptation of existing factors, and becoming as we know it today.

There are seven distinct floral regions of the land and sea, and while the tropics cannot be defined precisely with regard to natural biological phenomena, the flora which extends over the strictly intertropical zone and its neighbouring areas is grouped into four of these regions.

These seven floristic regions are the Holarctic, Palaeotropical, Neotropical, Cape, Australian, Antarctic and Oceanic, and it is the second, third, fourth and fifth of these which are associated with tropical flora.

The *Palaeotropical Kingdom* includes the flora of tropical Africa from the Sahara to the Kalahari desert, Arabia, India, Indonesia and the islands of the Pacific Ocean. Typical of this flora are the plants belonging to the Pandanaceae, Liliaceae, Urticaceae, Myrtaceae and Sterculiaceae families.

The *Neotropical Kingdom* includes the tropical regions of America, including Mexico and the Caribbean and the whole of South America, with the exception of Chile and Patagonia which form part of the Antarctic Kingdom. Neotropical flora is characterized by an abundance of Pontederiaceae (water hyacinths), Bromeliaceae (pineapple, etc), Cannaceae, Monimiaceae, Erythroxylaceae (including the coca), Passifloraceae and Cactaceae – the characteristic 'fleshy plants'.

The *Cape Kingdom* covers only the southern tip of Africa. Territorially, therefore, it is small, but its flora is far from sparse since it includes the following families, represented by a great number of endemic species: Iridaceae (including

The exotic bloom of Protea laurifolia, a plant found in southern Africa. An encircling halo of leaves protects the flower

15

many gladioli), Ericaceae (with about 500 species of Erica), Mesembryanthemaceae, Oxalidaceae, Geraniaceae (including many pelargoniums), Myrothamnaceae, Bruniaceae and Penaceaceae.

Finally, the *Australian Kingdom* includes Australia and Tasmania; New Zealand does not come within its limits as it forms part of the Antarctic Kingdom. The plants of the Australian Kingdom are very characteristic, and of the greatest interest are the species in the genera *Eucalyptus* and *Casuarina*, which are found beside representatives of the Proteaceae, Tremandraceae, Stackhousiaceae and Dilleniaceae families. The isolation of genera or families in any one of the floristic kingdoms, or in certain of their territories, is a result of the separation of the continental masses in the remote past which prevented any transfer of plants.

The Gramineae illustrate well the exot exuberance found in tropical vegetation. Typical examples, apa from the bamboos, are species of Cortaderia, Saccharum and Sorghum

The virgin forests

'Virgin' or natural forests conjure up an image of dense woodland extending over vast areas, with immensely tall and upright trees, their foliage overshadowing a wealth of undergrowth made up of ferns· and orchids, lianas, and unusual plants which harbour dangerous insects. Poisonous snakes slither along the ground and there is a background of screeching monkeys and parrots. This image of the jungle is conjured up in tales of adventure and the records of explorers, and is the jungle presented by Technicolor documentary films, and the travellers' tales so often seen on television.

The true picture is not so very far removed from this, but equatorial forests vary enormously; a more appropriate term is in fact tropical forests, since virtually identical plant communities occur in regions far removed from the equatorial belt. These lush tropical forests are replaced under different climatic and topographical conditions by park forests or littoral (riparian) forests, or by the semi-aquatic forests known as mangrove swamps.

Virgin forests are to be found in the regions each side of the Equator, in West Africa, in the southern part of Central America (particularly in Amazonia), and also in the Antilles, Jamaica, and a narrow coastal strip along the Atlantic shores of Brazil. There are also lesser examples in the eastern part of Madagascar and certain areas of India and Ceylon, Indo-China, and in particular in the large islands of Indonesia and the Philippines.

These forests mostly consist of evergreen trees – plants that do not shed their leaves seasonally – and they are subjected to a very hot and rainy climate with an annual rainfall of over 80 inches. For this reason they are also known as tropical rain forests.

The average temperature of the equatorial climate in which these forests grow varies from 24° to 35°C (75° to 95°F) with a minimum temperature of 18°C (64°F). The temperature range is not as extreme as one might imagine, and it is therefore possible for a great number of plant species to develop and flourish.

Nevertheless, a true equatorial climate differs from a tropical climate in the limited sense of the word, in that the latter is less damp, with its rainfall concentrated in fairly sharply defined districts. The length of the dry season varies, especially in those forests furthest from the Equator.

Knowledge of the structure, floral composition, and growth of tropical forests is still incomplete and has been acquired in relatively recent times. At the time of some of the earliest journeys of exploration, a French naturalist wrote: 'The vegetation of tropical Africa is little known to us as yet because of the terrible and insalubrious nature of those countries.

'Tropical vegetation is not uniform in its distribution, although many forms that are dominant in Africa are to be found elsewhere. In addition, some plant species that are normally herbaceous in non-tropical countries take on a more woody appearance in hotter and damper climates.'

This was a simplified view, but his observations show that about a hundred years ago the Brassicaceae, or Cruciferae family, and the Caryophyllaceae were not to be found in Equatorial forests, which is not the case today, even though they are more common in the milder regions of the northern hemisphere.

Systematic exploration, backed up by reports and photography, has been conducted since the nineteenth century, and much more information is now available. This exploration has served not only to penetrate the aura of mystery which used to surround the rain forests but also to throw light on their ecology, climate and flora. Although these factors are often similar, the equatorial forests vary in the composition of

their flora, according to whether they are locate
on continents or on large islands.

The environment shows a high humidit
because of the frequence of violent rainstorm:
These usually occur early in the afternoor
drenching the soil and the foliage, which cor
tinues to drip on to the undergrowth. With suc
an abundance of water, the atmosphere, too,
exceptionally hot and moist.

The soil is rich with nutrients derived from th
continuous natural supply of leaves, branche
flowers and fruits, as well as from the excremer
and remains of animals that have died there ov
thousands of years. All this material is fermente
broken down or transformed by bacteria an
fungi and provides the roots of dwarf an
vigorous plants with the rich humus required fc
good growth.

Vegetable life is also stimulated by the proce
of transpiration: evaporation of water from tl
leaves forces the sap to rise continually from tl
roots upwards, contributing to the establis
ment of an increasingly luxuriant vegetati
cycle.

Light is not as abundant as moisture in the
forests, or at least in the forests that are mo
typical in structure. Very little sun filters throug
the topmost branches of the tall trees that form
dense canopy of foliage. The environment
usually greenish and dim, forcing the plants
grow upwards rather than outwards in order th
their leaves may reach the light. This is the reasc
why the trees in the rain forests grow to heights
120 to 275 feet (40 to 79 metres), their long trun
forming incredible pillars with an intricate cor
plex of branches, liana-like ropes, and overhe:
roots hanging down to the ground like slende
immobile serpents.

Only in the few places where the plants th
out or the branches of the trees become le
twined, or where their leaves are smaller or mo

Left: A hanging forest in the Fiji islands. The moisture from
hidden watercourse under the treetops produces this luxurian
Facing page: An impenetrable curtain of ferns and liana in
Fijian virgin forest

divided, does a slightly brighter light reach the undergrowth. One effect of light shortage can be observed in the case of plants with wide glossy leaves, a feature that has earned them the name of 'tin leaves'. These reflect the beams of light downwards, like oblique mirrors.

The trees in these forests can reach heights of up to 275 feet (70 metres), but heights of 120 to 140 feet (40 to 50 metres) are more common. Although there are open clearings here and there, where the foliage forms a dome above, the vegetation is generally mixed and untidy: it falls to the ground in tangles or hangs down like huge curtains of living material. It often seems to be built artificially and may be studded with gaudy flowers such as orchids and Tillandsia, and with

Above: Tropical forest in New Guinea. Facing page: A study in textures and shapes: 'Fern Valley', Hawaii

the green fronds of ferns and other epiphytes.

In these forests, representatives of very ancient groups of plants are found growing side by side with recent and more specialized plants. Hence the co-existence of arborescent ferns, Lycopodium, Selaginella and Cycadaceae, (relics of Carboniferous landscapes) in close proximity to families such as the Scrophulariaceae, Malvaceae, and many others which can unhesitatingly be classified as belonging to much more recent groups.

The tree composition, too, is heterogeneous, with large Lythraceae, Bignoniaceae, and Verbenaceae, and trees which produce highly prized hardwoods such as teak, as well as species from the genera *Ficus, Sideroxylon* (another iron wood), *Canarium*, and others. Many species of plants in these families are the equivalent of herbaceous plants in more temperate climates, or at least of bushes and thickets.

The rain forest also contains species of which there is no representative of the same genus or even of the same family outside the forest environment. Sometimes, however, specific plants are found as fossil relics in widely separated territories: for example, leaves and other fossil remains of the Lauraceae genera (such as *Cinnamomum*) which are now to be found only in the tropics, have been discovered in fossil form as far away as the subsoil of Greenland.

The strangest forms of plant life exist in these forests: for instance there are large trees whose foliage is so extensive and heavy that natural buttresses are needed to keep them upright, trunks which are contorted into shapes such as bottles, or which sprout forth thin branches at every point, and trees whose trunks are garnished with sessile flowers and fruits, as is found in the cocoa tree, a typical inhabitant of the tropics.

The roots hanging down to the ground from the treetops are known as aerial roots, and are characteristic of such trees as the pagoda *Ficus*

A fig tree on the edge of virgin forest in New Caledonia. Its elegant supporting roots hang from the branches

(*F. elastica, F. religiosa, F. aurea*). They usually form an intricate and remarkable 'elephant paw' configuration, often merging together as they grow and producing monstrous formations as they become attached to the colossal trunk.

Other strange plants found in the rain forests are the many parasites and semi-parasites. Some live with hardly any chlorophyll at all, others have a very limited amount of this pigment. The parasite species are usually yellow, ochre or reddish in colour; they are sometimes strange in appearance, with beautiful red or gold flowers,

and they can live even in areas of minimal light, as they are able to obtain food independent of the process of photosynthesis. Indeed, these curious plants live entirely off their host.

The semi-parasite plants, on the other hand, attach themselves to the roots or branches of the host plants by means of pads or special sucker organs (like the mistletoe of temperate climates); *Loranthus* is a good example, and is abundant in the damp forests of the Equator, while *Loranthus europaeus* has its main centre of distribution in Asia Minor.

Above: A glade in the vast forest on the slope of Kilimanjaro (Tanganyika). Right: Many species of tropical trees are anchored to the ground in a strange and tortuous way. This picture shows the root of Scolopia crassipes (Flacourtiaceae) from the wooded regions of India and China

Park forests

The term 'gallery forest' or 'park forest', as we have said, describes a very different type of tropical vegetation from the natural jungle that is the virgin forest. For unlike the virgin forests, the park forest, as its name suggests, is relatively easy to penetrate, often with light airy clearings, and has vegetation which is considerably less luxuriant than that characteristic of the true lowland rain forest.

Typically, a park forest consists of tall trees forming a vault-like covering over low under-growth and brush. It is most usually found along the banks of a river, winding along its course for many miles, and stretching out on either side to cover a considerable acreage. Park forests are often crossed by innumerable tracks with which the local inhabitants are extremely familiar, but which are almost impossible for anyone else to detect. Here and there are groups of palms, those typical of the region in question, or of the continent itself: *Raphia* in Africa, for example, and *Euterpe* in Brazil. Apart from the palms,

Right: A group of gigantic close-packed bamboo (Dendrocalamus giganteus) in a Ceylon forest. The stalks of these strange Gramineae may sometimes grow to the size of a man's body

Left: The variety of shapes assumed by tropical plant species is infinite. This tangle of contorted branches is part of a forest on the island of Oahu, Hawaii

large bamboo thickets are very common, especially on river banks and around the edges of clearings.

Despite the apparent homogeneity of these forests, it is possible to distinguish different types of vegetation. The Cyperaceae, Araceae, Liliaceae and Bromeliaceae abound in the forests of the American continent, while in the Indonesian and Australian forests Begoniaceae, Urticaceae, Zingiberaceae, Ferns and Selaginellas are more common.

In these forests, which follow the rivers and which sometimes appear to have been created by man, like an artificial park, there are very few trees of truly gigantic size, and the clearings with most light can be seen from far away. In places where there is a clearing, the vegetation changes perceptibly. Low green tangled undergrowth gives way to tall dry grasses, particularly species of Gramineae, which grow together closely in the spaces between the trees. In these clearings we encounter vegetation that is reminiscent of the thickets and savannas that border the great tropical forests.

The palms

Most people are familiar with two types of palm: the date and the coconut. These are found in two different environments, the first a native of the oases that give life to the deserts of North Africa and Arabia, the second characteristic of beaches and coral reefs, especially those of the Pacific Ocean.

The palms are considered symbolic of the tropics more than any other group of plants. Except for a dozen species that grow in the temperate zones of the two hemispheres, all the others are characteristically tropical flora. They number more than a thousand species, which are native to Malaysia, Amazonia and, in a few cases, Africa.

Despite their apparent affinity with a single environment – the hot desert – palms grow naturally in a very wide range of locations. Some, like the date palm (*Phoenix dactylifera*) and the *Hyphaene* thrive in the hot arid climate of the Sahara and Arabian tropical deserts. They flourish around oases where water comes to the surface, as they need a certain amount of moisture in the subsoil.

Other species of palm grow in the savannas to form large and fairly homogeneous populations, often withstanding the aridity of the dry season and even the fires which destroy vast areas of high grass.

The palm dominates, however, in the damp tropical forest where it is distributed according to its special characteristics. In some cases, palms are isolated amidst the green of the dense lower vegetation, in others they form close-set groups near swamps, as in Amazonia.

Most palms provide an important source of materials, especially foodstuffs. For the local population, palms have always formed an important part of the food economy, and many species provide food for the civilized world – and so a source of income for the growers. In the case of the sugar palm a flow of sweet sap is provided when incisions are made in the stemapex, or the inflorescence is decapitated, Unfortunately, such methods usually lead to death of the plant. The liquids are very often employed not only for the extraction of sugar but also to produce an excellent palm wine by fermentation. Palm wine is, indeed, a drink of some antiquity: as early as the fifth century BC, the Greek writer Xenophon told of the dire effects of the wine on the Ten Thousand, retreating from Persia.

Among the sugar-producing species, of special note are *Arenga saccharifera* of tropical Asia and Indonesia, which also provides good quality textile fibre, *Borassus flabellifer* from West Africa and India, and *Nipa fruticans*, common along the coasts of the Far East.

Facing page: Coconut palms along the beach at Moorea (Polynesia). These trees are widely distributed along many equatorial coasts. Right: Sword-like palm leaves, Hawaii

Of great importance too are the oil palms, including *Elaeis guineensis*, which lives in the secondary forests of the whole region stretching from Senegal to the Congo, and the *Orbignya martiana* in certain Brazilian forests producing nuts from which a high-grade industrial oil is manufactured.

Among the palms that supply fatty substances, *Copernicia cerifera* should not be forgotten. This is the wax palm which lives in the dry northeast sector of Brazil. From the surface of its wide flagellate leaves is scraped the 'carnauba' which is a source of wealth in these regions; the seeds of the palm are also used locally as a coffee substitute.

The most important of the whole palm family is without doubt the coconut tree (*Cocos nuci-fera*), probably originally a native of Indonesia, which has spread along most shores of the tropical seas, its large fruits being transported there by ocean currents. Every part of this plant is put to good use, from the timber of its stem to the long leaves that thatch the roofs of native huts and the fruits that provide both food and drink in the form of the liquid 'milk' of their seeds, the coconut. Fats and oils are also derived from this palm.

The 'vegetable ivory' (*Phytelephas macrocarpa* and *P. microcarpa*) and the 'sago' palm (*Metroxylon rumphii* and *M. sagu*), are also very useful members of the palm species. The pith of the stems of the sago palm produces a starchy flour that feeds entire populations in the African and Asian tropics.

The Antilles: Oreodox oleracea produces young shoots that are edible and are commonly known as 'palm cabbages'

Lianas and ferns

The lianas that hang down from the trees like woody cords and the tree ferns have already been mentioned. In addition, there are other ferns, termed *epiphytic*, which do not live on the ground but are attached to the bark of trees, sometimes taking root there.

Lianas are essentially twining plants. They require little light when first growing, but as they reach maturity, they are forced to climb upwards towards the sun. In order to do this, they have to attach themselves to sturdier trees, for they are not self-supporting, owing to the slender diameter of their stems. They cling to their hosts by means of tendrils similar to those of the vine or the pea, and in many cases there are sucker pads at the tips, as in the case of the virginia creeper. Lianas are largely responsible for the dense and luxuriant appearance of the rain forests and are virtually non-existent in more temperate climates. Certain ferns in the genera *Trichomanes*, *Lygodium*, and *Lomariopsis*, abundant in the dampest parts of the forests, are very similar in habit to the lianas.

Most lianas are dicotyledons, and a relatively few belong to the monocotyledons. Some lianas are herbaceous, others woody, and still others are palms in the genus *Calamus*, including the 'rotang', from which long, highly flexible rushes – ratten cane – are derived. They live in the forests of the Old World, where there are about a hundred species, several of which reach lengths of 500 to 600 feet. The slender stems of these curious palms do not rise vertically but twist and turn, keeping to the middle layer of the forest.

Many lianas belonging to the dicotyledons (Leguminosae, Convolvulaceae, Cucurbitaceae, Piperaceae, Vitaceae, Passifloraceae, Cyclanthaceae, Sapindaceae, and Polygalaceae families) have more rigid, larger stems. They are usually as thick as a man's arm in diameter, and sometimes as thick as a man's body, although they are always disproportionately long. They are generally circular in section, although flattened types, simulating a dorso-ventral structure, can also be found.

Bare stems may be seen, hanging straight like large tent ropes, encased by other thickened or tapering lianas interwoven in an intricate pattern and bearing tufts of leaves, flowers, and fruits.

Man has managed to find a use for some of these plants: he uses the straw from the leaves of the *Carludovica palmata* to make Panama hats.

A liana, resembling twisted cords, in the New Guinea jungle

Left: A common
curiosity of tropical
plant species is the
'aerial' roots, which
hang from the branches
like ropes (the plant
shown here is a
Bauhinia from India)
Right: The twisting
trunks of Phoenix
reclinata, a date palm
from the forests of
Ceylon

This liana belongs to the Cyclanthaceae family and is native to the forests of the Andes in Ecuador. The fruits of certain American Passifloraceae are edible, while those of the Amazonian *Paullinia cupana*, a member of the Sapindaceae family, are used in the preparation of 'guarana', a popular Brazilian cocoa-type drink. The African liana *Strophanthus gratus* furnishes a medicinal alkaloid drug, strophanthin, which has an effect on the heart similar to that of digitalis. Finally, the Chinese yam *Dioscorea batatas*, also a liana, has tubers containing an abundance of starch, as does *D. alata*, the white yam from the Himalayas, which was introduced into America in the middle of the sixteenth century.

Most experts agree that some of the strangest

Facing page: Examples of epiphytic plants from equatorial forests: an imposing cascade of ferns on a palm shoot (top); Platycerium, the Elkshorn fern from New Guinea (bottom left); Ophioglossum pendulum from the Ceylon forests (bottom right)

Right: A forest on the slopes of Kilimanjaro (Tanganyika). Mosses, ferns and other parasitic plants cover the trunks and branches of the older trees

lianas are those belonging to the genus *Bauhinia* (Leguminosae) found in the forests of South America. The curious feature of these lianas is their stem which has a regular succession of small flattened and large bulbous segments. The bark looks as if it were burnt, but it bears beautiful orange-coloured flowers.

Growing among this intricate confusion of rising and twisting stems are tufts of a delicate, tender green. These are the ferns, which find the humid heat of the forests an ideal habitat. It is not uncommon in the jungle to come upon tree trunks and palm stems completely covered with epiphytic ferns. This exuberant growth may be due to the accumulation of water and humus in the hollows formed at the base of the large leaf petioles that break the regular surface of the palms. Here the ferns, like many other epiphytes, can take root and find all the nutritional substances they need for their life-cycle.

Among the most flamboyant and common genera in the Indo-Malaysian and African forests is the fern *Platycerium* and above all the great *P. coronatum* of Malaysia, whose leaves often attain a length of more than six feet. A minor species is the *P. alcicorne*, or elkshorn fern, native to the Australian forests.

No less elegant are plants of *Asplenium nidus* with their long fronds, reminiscent of rigid banana leaves, and a large black and glossy rachis in the middle. Despite their sometimes colossal size, these ferns manage to get inside the trunk of many species of tree, where they hang like large baskets.

Above: The twisted branches of the banyan tree, Ficus bengalensis, found in the forests around Colombo.
Right: Palm-like fern (Dicksonia antarctica) natives of Australian and Tasmanian forests.

Orchids and other flowers

Many flowering plants are found among the epiphytic flora. Research on them has shown that each epiphyte will gravitate towards certain plant species in preference to others: in other words, it has some affinity for its host, although it merely attaches itself on the rough surface of the bark or branches without in any way damaging the plants that give it shelter. Epiphytes are not parasitic, as they take no nutrients from the plants to which they are attached. They manufacture their food in exactly the same way that other plants do, but instead of obtaining their moisture from the earth, they take it from the air that surrounds them.

The outstanding epiphytic flowering plants include the *Tillandsia* (one of the Bromeliaceae) from tropical America which produces long

pendulous growths or 'beards'. Its festoon-like appearance is so striking that one species, common to Cuba, has been called *Tillandsia usneoides* (popularly known as Spanish moss or old man's beard) because of its similarity to *Usnea barbata*, a common lichen that is draped from the branches of firs and larches in the forest regions of the Northern Hemisphere. The silver-grey *Tillandsia* plants, with their long, narrow leaves and small yellow or reddish flowers, have adapted so well to a 'pendant', rather than a merely epiphytic way of life that large amorphous gray masses are often seen growing on electricity cables, their seeds having been carried there by the wind and birds. Other species from the same genus have become adapted to the driest climates, as is demonstrated by those which take root on large Cactaceae, forming tangled skeins. There are many examples of *Tillandsia* living in desert areas, even on the ground.

Botanical families that contain examples of one or more epiphytic plants include the Begoniaceae, Melastromaceae, Gesneriaceae, Liliaceae, and even certain Cactaceae. However, the Orchidaceae undoubtedly hold the place of honour for the beauty and abundance of their flowers, which flourish on branches and trunks. A single flower unfolds from a mass of leaves, or a dainty cascade of tiny blossoms falls from its niche between the fork of two branches.

The flowers are to be found in almost all colours, and indeed there may be as many as six or seven different shades in a single blossom. Despite the widely held belief that there is a wholly black orchid, the nearest colour is in fact a deep brownish purple. The rarest colour is blue, although there are about twenty species bearing flowers of differing blues.

Two examples of tropical orchids.
Left: A Cymbidium. Right: Vanda hybrida

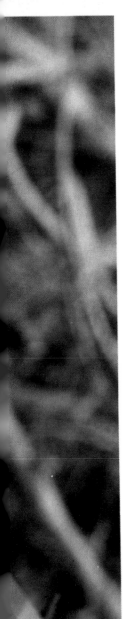

Above: Odontoglossum cirrhosum from Ecuador.
Right: A Miltonia

Facing page: A selection of the many decorative exotic
orchids. Far left, from top to bottom: A Cattleya, Odontioda
cooksoniae (a bigeneric hybrid), and Huntleya meleagris.
Near left, top: Epidendrum prismatocarpum from Panama.
Near left, bottom: A Cymbidium

Orchid flowers are unmistakable: they are irregular, being made up of six sepals or of three petaloid sepals and three petals, the lower of the three, the labellum or lip, jutting forward. The shape of this labellum varies greatly and may be very strange: in some cases it is laminar and fringed, in others it is merely a laciniate fan, and in yet others it is bulbous, pouched slipper, pipe, or trumpet-shaped.

There are approximately 20,000 species of orchid in some 700 genera, and they are to be found in every climate. Four-fifths of them live in tropical and subtropical zones, where there are no doubt a great many species as yet undiscovered. It is in these areas that the epiphytic types predominate.

The countries with the greatest number of orchids are New Guinea (Indonesia), with more than 2,000 species, Costa Rica, the Philippines, and Venezuela, with species in the genera *Masdevallia*, *Miltonia*, *Odontoglossum*, *Paphiopedilum*, *Cattleya*, and many others.

Even though rain forests are so uniform, the distribution of epiphytic orchids is very individual: most prefer places where there is light, but a minority tend to take root in shady nooks where the air is damp and relatively cool. The *Phalaenopsis violacea*, for example, grows in moist, dark woods in the valleys of Sumatra. Others like to live on higher ground, in scrubland, and they often adapt to a terrestrial rather than an epiphytic existence.

For various reasons orchids, like thousands of other plants species, are becoming more and more rare, even in their countries of origin. Man is a collector and has long established a highly profitable trade in these plants. One day many species will be no more than a memory, made familiar by specimens pressed between the leaves of a collector's book or by pictures in a photographic library, or at best a rarity to be found only in botanical gardens.

In order to appreciate the beauty of these flowers, some specific examples are necessary. The plant *Stanhopea tigrina* from Mexico is regarded as the most splendid of all the Stan-

Above: A group of highly decorative orchids, including the beautiful Phalaenopsis leslie, from Ceylon (foreground). Left: Flowers of a New Zealand Cattleya

hopeae. This orchid is remarkable not only for its size, shape, and colouring, but also for the thickness, texture, and fleshiness of its sepals. Its fragrance, reminiscent of vanilla, is so strong that it is overpowering indoors; while in its natural surroundings, it is noted over a wide area. Its flowers – at least in its country of origin, if not in conservatories – are seldom single; each inflorescence may include two, three, or even four flowers. The wide wax-like sepals are yellow, sometimes tinged with green, or sometimes honey-coloured, with irregular stains of dark purple. The labellum is perhaps the most complex part of this flower: it is recessed and projects forward with three pointed laminae, also yellow in colour and spotted inside with carmine.

A species in the same genus, *Stanhopea costaricensis*, is a native of Costa Rica, as its name implies. It is a little less extraordinary than *S. tigrina*, but it is remarkable for its shape. Its sepals are ivory-coloured and tinged with green, or a tender yellow with purple-black splashes; some of the markings are large, some are small, and many of them have 'eyes'. In this species, too, the labellum is hollow and curiously shaped, with a large projecting pincer whose lower edge has two symmetrical side appendages shaped like horns, features that have given the *Stanhopea costaricensis* the local name of 'torito', or 'little bull'.

Less showy but just as decorative is the *Stanhopea ecornuta* from Central America. The specific epithet is due to a peculiarity of the labellum, which is fleshy and dilated, but with-

out the horn so characteristic of most of the *Stanhopea* genus. It is wax-like and shiny, yellow-gold near the base and white or almost white at the apex, and dotted with purple. In this species, the flowers almost always come out in pairs.

The most regal and fragrant specimen of the orchid family is the *Stanhopea tigrina*, but the most breathtakingly beautiful is perhaps the *Laeliocattleya callistoglossa*, which is a hybrid of the *Cattleya warscewiczii* from Columbia and the *Laelia purpurata* from Brazil. It is a sizeable lilac-pink orchid with large, fringed sepals and labellum. It has been bred by floriculturists after years of work on selection and hybridization, and reflects the brilliance of its two parent species.

The beauty of the *Cattleya* and *Laelia* flowers is familiar not only to the botanists who first encountered them in the thick equatorial forests, and to growers of these rare species, but to all florists and lovers of exotic flora. The *Cattleya dowiana* from Columbia and Costa Rica, for example, produces large flowers, their light yellow sepals contrasting with a wide trumpet-shaped purple labellum with cockled edges. Equally beautiful is the pink and lilac *Laelia gouldiana* from Mexico with its three-lobed labellum.

The flowers of many orchids may be more than strange: they may be grotesque, as though from another world. The *Brassavola digbyana* from Yucatan and Honduras is a very good example: it has one large, sweet-smelling flower, with six fairly similar long sepals arranged in a star. In colour these are a milky greenish-white and from them protrudes a large, roughly circular-shaped labellum with three lobes, each of which is edged with a deep and fine fringe.

Even stranger are the flowers of certain *Cirrhopetalum*, especially the *C. medusa*, commonly known as the 'Medusa head orchid', which flowers in autumn in the forests of the Sunda Islands. The whitish flowers are freckled with red and are grouped in large bunches towards the apex of a common peduncle. While their upper sepals are short, lanceolate, and pointed, the

46

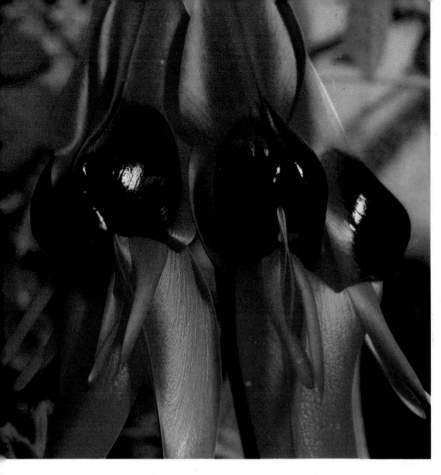

lower sepals are far longer and thinner, and this particular structure gives the orchid a wig-like appearance.

In *Cirrhopetalum fascinator*, a native of the Annam forests, the flowers may be paired or solitary and are fairly gaudy. They have three upper sepals or a rounded triangular shape with an elegant long fringe all round; the lamina is white with delicate yellow and greenish tints and it has five stripes speckled with dark purple. The surrounding fringes are violet and pink. The

Facing page: Two magnificent and unusual orchids: the highly perfumed Stanhopea tigrina from America (top), and the unique Dendrobium ratiotes, found in tropical Asiatic forests (bottom). This page, above: Clianthus ampieri from Australia. Right: Stapelia variegata, the carrion flower, a South African asclepiadacea with large, russet-coloured flowers

two lower sepals (the lateral sepals), are of a similar colour and also have longitudinal stripes, but they differ in that their sealed edges part at a certain point, and hang down like long slender tails, often more than 30 centimetres (about a foot) in length, twisting round each other at the bottom in a spiral shape. From the middle of the upper portion of these two sepals grows a small, yellow, tongue-shaped labellum.

The flowers of certain catasetums are even more unusual in appearance. They include the *C. fimbriatum* of the Argentinian and Venezuelan forests. Their sepals are lanceolate and rigid, and most of them are curled inwards all the way down their length. They are pale green suffused with pink in colour. The labellum is large, rounded, and shell-shaped; it too is pale green, and of a fleshy consistency, surrounded by a coarse fringe.

Another species of the same genus, the *Catasetum gnomus* from Amazonia, has large and strange flowers with the two upper lateral sepals almost concealing the greenish, violet-tinged upper sepal. The two lower side sepals are long and purple in colour, curling and stretched apart like symmetrical wings, while the large, cavernous, orange lip has a rounded hood, with toothed edge, its pure white border folding backwards.

From a different genus, the *Coryanthes albertinae* of the Venezuelan forests has purple spotted flowers dominated by a large open-mouthed concave labellum; the sepals retract as the flower opens out. A biological peculiarity of this flower is that above the labellum is a columnar organ with two appendages on either side, which are thought to act like a trap. These secrete drops of a clear liquid into the bowl-like cavity of the labellum which slowly fills up to the brim, attracting large pollinating insects that slither into the container. In their violent struggle to escape from the trap, the insects are thrown against the pollen which adheres to their bodies and is subsequently transported to other flowers for fertilization. In this way, perpetuation of the species is ensured.

Above left: A group of epiphytic orchids from Ceylon. The characteristic aerial roots may be seen between the leaves.
Above right: Passiflora racemosa flowers from the forests of America

Rivers, marshes, and mangrove swamps

The term mangrove swamp is applied to communities of tropical aquatic and subaquatic vegetation on the banks and flood plains of rivers, beside stagnant lagoons, and bordering the shores of estuaries and coastal areas.

The oddest feature of this tropical vegetation, which some botanists have called 'amphibious', is the small mangrove forests of hydrophytes growing in muddy swamps. These mangroves are formed by species of the genus *Rhizophora* and the few allied species which stretch out along all the coasts. Whether the mangrove forest is situated beside a low, tide-washed tropical sea coast or whether it is beside a brackish lagoon extending for miles along an estuary, it is always rather sombre in appearance.

The mangrove forest, or 'mangrovia', is typical of the equatorial regions of Africa, Madagascar, Indo-Malaysia and especially of certain areas of Central and South America (Guyana, Amazonia). More rarely, it is to be found on the coast of Japan, New Zealand, the Red Sea (Gulf of Aqaba), and even in the Somali Republic, where, for want of better fodder, the camels browse on the leathery, grayish leaves of the plants.

From the ecological point of view, this vegetation can be divided into two types. The first becomes established in saline areas some distance from free-running water. It consists of trees and shrubs which are able to withstand the high salt content of the substratum, a condition which makes for rigorous preselection since relatively few plant species are halophilic (salt-loving). Among these is the *Nipa fruticans*, a Malaysian palm which accumulates a high percentage of sodium chloride; indeed, the local people reduce the trees to ash, which they then use instead of mineral salt for flavouring their food.

The second group of plants is at home in places where there is a higher water table but it is less stagnant, and has a lower salt content. The plants extend their many roots down into the swamp, where the mud contains a wealth of organic material. There, deep in the mud, a special microflora produces intensive and continuous anaerobic fermentation. It is very difficult for the roots of the plants to breathe in anaerobic conditions (absence of oxygen), a handicap that is aggravated by the high surrounding temperature. The flow of oxygen from the air to the roots is insufficient for the metabolism of the root tissues, and the plants have developed additional means of survival. They develop woody 'pneumatophores', or special respiratory roots, with 'pneumathodes' or lenticel-like openings, and these branch roots rise vertically from the submerged roots to break the surface of the water, even during high tides. Through the open pores, they can absorb oxygen from the air to enable the root apparatus and the whole plant to live and flourish.

Many respiratory roots also grow downwards from the tree trunks, or branches producing a dense mass of tissue. During low tide, these are almost completely exposed and form an intricate scaffolding, giving the forest its characteristic appearance of giant spiders or plants on stilts.

Even though the balance has apparently been restored by the providence of Nature, plants still have to struggle for survival in these swamps. Unless special measures are available to prevent the seeds from dying of asphyxiation, while buried in the fermenting mud, these curious plant associations will find germination difficult. Oxygen must be available and, in fact, the seeds of *Rhizophora* and *Bruguiera* – the plants that make up the mangrove swamps – can germinate simply on contact with the humid air, without having to be buried in the watery substratum. They germinate while they are still on the twig, and only when the new root has grown to a foot or so in length do the seeds become detached from the parent plant. They fall and become embedded in the mud below. This process

ensures regeneration: ungerminated and unprotected seeds would almost certainly be washed away by floods or suffocated by the mud. In some other species, the seeds are encased in a thick spongy tissue which enables them to float for long periods. Carried along by the current, they eventually germinate when they are caught up in some obstacle. Only under these circumstances can they put down the roots that will ensure their survival.

In appearance, mangroves look like shrubby thickets 8 to 20 metres (26 to 65 feet) in height, made up of fronds bearing hard leathery grayish leaves and undistinguished flowers. Fish and other water animals of the genera *Periophthalmus*, *Uca*, and *Birgus* teem around the roots of the swamp vegetation, and are occasionally able to leave the water and remain on the banks for long periods. Indeed, mangrove swamps house a

multitude of amphibious animals and insects, because the stagnant water, rich in organic substances, is very suitable for animal life.

Apart from the *Rhizophora*, the palms of the species *Nipa fruticans*, and *Bruguiera* (Rhizophoraceae), mangrove swamps include species of the genus *Sonneratia* (Sonneratiaceae) closely allied to the Lythraceae (with which they were once grouped), and *Avicennia* (Verbenaceae).

Soft-water marshes in inland areas in the tropics contain a very different flora, the difference lying mainly in the wealth of the plant species. These belong to a wide range of families: the Gramineae, including certain bamboos and *Phragmites communis* (a common reed from temperate and marshy areas), the papyrus and the narrow-leaved reed mace (*Typha angustifolia*), the *Ravenala* and the Araceae. Even conifers such as the *Taxodium distichum* are included.

Riverside vegetation. Above: In a palm fore[st] in Marawila, Ceylon. Right: Through fores[t] and glades at Kauai, Hawaii

This species of cypress develops 'knees' which project from the root system to above the water level, and is typical of the Florida swamps.

The *Ravenala madagascariensis* has already been mentioned: it is commonly known as the 'traveller's tree', and is a spectacular member of the banana family. It can grow to heights of up to 30 metres (100 feet), and has a plume of large fan-shaped leaves at the top. The thick petioles retain water, and thirsty travellers can break them off and suck out a quantity of delicious cool rainwater from the base.

Notable among the Araceae of the tropical swamps are the *Montrichardia arborescens* from Brazil and the *Typhonodorum madagascariense* from Madagascar. The latter has large leaves very similar to those of the *Colocasia* or West Indian kale which are grown in almost all temperate regions. Even palms are to be found in swampy areas, particularly species from the genera *Euterpe* and *Raphia*.

The papyrus (*Cyperus papyrus*) inhabits marshes and the banks of the great rivers such as the Nile, up to the Belgian Congo and Lake Chad. Other smaller Cyperaceae, similar to large sedges and bulrushes, crowd together in compact colonies – floating islands, or natural rafts formed by the accumulation of miscellaneous debris including branches and tree trunks.

Aquatic flora brightens the immense stagnant pools in the tropics, and the wide, slow-moving rivers like the Amazon and the Nile, with large and brilliantly coloured flowers.

The Amazon is the home of *Victoria amazonica*, the royal water lily whose circular leaves may be two metres (six feet) in diameter, and can support the weight of a child. Its flowers are very similar

to those of the common water lily, but larger in size and are carmine pink in colour. The floating *Eichornia* and *Pontederia* are also to be found in this region, and they strike a note of singular beauty in the landscape.

Eichornia azurea and *E. speciosa* – otherwise known as water hyacinths – float quite freely and have a long tuft of fine branch roots which hangs down in the water. Leaf petioles spread from the collar in all directions; their edges are leathery, kidney-shaped, and concave. The bladder-like petioles with their very light white core give the plants their buoyancy. From the centre of the rosette of strange leaves rises a scape about twenty centimetres (nine inches) high; this has a beautiful erect panicle consisting of violet-blue flowers, each of which has six sepals.

Very different is the lotus, with its wide, almost circular, umbilicate leaves, opaque and of a gray-green colour, which grow on long petioles rising from the water. The flowers, too, are large, attaining a diameter of six to twelve inches (up to 30 centimetres). They are formed of many pink petals around a crown of yellow stamens, which in turn are grouped around the ovary that later turns into a fruit. The fruit is strange, with a cup as large as a fist, a convex sealed top, and many perforations in each of which is concealed a seed.

These seeds, known as Pythagoras beans or Egyptian beans, are edible and floury, tasting of aniseed and containing starch.

A related species (*Nelumbo speciosum*) is a native of India and produces pink or yellow flowers. Both species of lotus have now been introduced into almost every hot or temperate region and are cultivated for their decorative beauty.

Thus, although we have said that mangroves and similar environments are not of spectacular beauty, and although it is true that the contrast between them and the tropical forests is striking, nevertheless the wet, swampy habitat harbours a number of lovely plant species, and the whole view of riverside greenery is often one that holds a bold splendour and lushness.

Having now considered the species which one tends to associate with the tropics – the trees of the jungle, the palms, orchids, ferns, mangroves and swamps – we shall move on to describe other habitats in which the temperature may still be high, but which present different problems for plant adaptation and survival: mountains, savannas and steppes, and deserts.

The marvels of tropical landscapes are not confined to the humid regions of jungle and river. There are, within the tropical zones of the world, great mountain ranges which can support a variety of vegetation, helping to give them their gaunt and majestic beauty. Technically, the best known of these, the Himalayas, lies above the intertropical belt, but many other familiar ranges are closer to the equator: for example, the ranges of Kenya, Kilimanjaro and Ruwenzori – all in Africa – and the Andes of South America.

It is these mountain environments and the vegetation that they support, which will be described and illustrated in the next chapter.

Above: The Kala Oya river (Ceylon), whose slow-flowing waters ar completely covered by vegetation. Below: Lotus flowers and leaves, in an African river

Mountain vegetation

In both tropical and temperate zones, environment alters drastically according to altitude. In the space of 200 metres (650 feet) up the sides of a valley or on the slopes of a mountain, great differences in plant communities can be seen.

Explorers say that high altitude tropical vegetation is at its most magnificent on the great mountains of equatorial Africa, where the lower mountain belt (often consisting of conifers similar to junipers, or of bamboos), gives way to very dense growth of arborescent or semi-arborescent *Erica*.

The beauty of these plants lies mainly in the spectacular mass flowering, when they are covered with innumerable quantities of minute white or pink blossoms; in other cases, the gray-green *Philippia* provides the dominant note of colour.

A similar arrangement of varying species characterizes all the large mountain ranges of Africa, showing not only considerable differences between one range and another but even between slopes of a single mountain.

The lower slopes of the Kilimanjaro, for

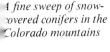
A fine sweep of snow-covered conifers in the Colorado mountains

example, are covered with savannas similar to those that stretch out from the foot of the mountain as far as the eye can see. The southern slopes, however, are clothed in the vegetation of a rainy, almost temperate climate.

On the west and east flanks of the Ruwenzori, opposite the Congo valley, the vegetation is that of a tropical rain forest. Above this level, the two mountain chains are both covered with conifer forests, of *Juniperus procera* and *Podocarpus*; there is also an area of epiphytes and lianas, and higher still up the mountain, a thick group of bamboo.

The mid-mountain vegetation of these rocky masses is again different. The vast damp meadows of Gramineae begin at a height of about 2,500 metres (about 8,000 feet); giant lobelias start to flourish at this altitude. Their large woody stems stand upright like bayonets with a plume of long leaves in a rosette shape, from the centre of which rises a columnar spike of beautiful flowers. The spectacular giant senecios (Compositae family) also grow here, sometimes reaching heights of 15 metres (50 feet).

These are the plants, particularly the *Senecio friesorum*, that grow at altitudes of up to 4,500 metres (13,000 feet) among vegetation that becomes increasingly sparse and stunted; at these high altitudes, dwarf species of the genus *Philippia* are to be found.

The *Senecio* that live at these altitudes have become adapted to the cold; such is the case of *S. friesorum*. Most species are equipped with protective leaves to form a sort of sleeve rising from the very centre of a rosette consisting of leaves of a different shape. Different forms of *Senecio* live at a height of about 450 metres, on Mount Kilimanjaro and in Kenya. One of the most spectacular of these is *S. johnstonii*, a tall, thick plant, sometimes ramified, with an enor-

*Semi-desert in flower in Colorado
at the beginning of summer*

*On the previous pages: A view of the Rocky Mountains. The
flower-studded meadows, wooded slopes and rocky peaks are
reminiscent of those of many mountain regions*

mous tuft of lanceolate leaves at the step top o
each of the few large deformed branches.

Other African mountains, such as the Semiel
range in Ethiopia, have a dominant and homo
geneous vegetation. Within the altitude range o
400 to 500 metres (around 1,500 feet), the plan
life may include the *Carex* (*Carex monostachys*
and among it grows *Lobelia rhynchopetala*, i
thick clusters of long, cylindrical conical spike
of flowers.

At an altitude of about 5,500 metres (18,00
feet) the lobelias and senecios thin out and gras
gives way to moss and sphagnum. The flora c
mountains is represented by minute herbaceou
species belonging to some of the same genera tha
populate the Alps, for example, the *Arabis* an
certain types of *Poa* and *Luzula*. Finally, abov
these, are the eternal snows.

The typical landscape of the slopes of Moun
Tsaratanana in Madagascar consists of forest
in which mosses and lichens predominate, form
ing a deep olive-green stratum; this graduall
gives way to trees that can be defined as dwar
since they do not usually exceed a height of 1
metres (50 feet).

Different again is the vegetation of the moun
tains of the American tropics. The base is ofte
cloaked with a mantle of *Podocarpus* forest
(conifers), while the belt immediately above th
has a sparser tree vegetation.

Further up, at around 4,000 metres (13,00
feet), where the weather is colder and drier, th
characteristic 'paramos' or 'waste lands' are to b
found. This unusual and desolate region i
populated by prostrate dwarf bushes, and i
steppe zones by Gramineae (genus *Calc
magrostis*). At 3,800 metres (12,400 feet), nume
ous species of Asteraceae or Compositae flourisl
sometimes reaching a height of three metre
(10 feet). An example is in the Sierra di Sant

*Left: Another view of the Colorado flowering semi-desert. It
is now no longer high season, and the colours are fading. Rig
Mountain vegetation in Nevada recalls an Alpine scene. The
species, however, are different, although they belong to the
same botanical families*

Left: Another Colorado scene. The exotic broadleaves resemble those of European poplars and birches, changing colour and losing their leaves at the beginning of autumn in the same way. The conifers keep their dark green foliage.
Above: Calotropis procera in fruit in a mountain range of the Algerian Sahara.
Right: A collection of Senecio species, typical of the high Kenyan mountains

63

Left: A striking group of lobelias, growing like a tree in Kenya.
Below: An aster similar to the European Aster alpinus, found on the remote moorlands of America in spring

Domingo in Venezuela. There are also lovely Bromeliaceae of the genus *Pourratia* with spinescent leaves and bushes bearing small red flowers that glow like flames, known locally as 'chuquiragua'.

In the Andes, plant life grows even higher than 4,000 metres (13,000 feet). The rare, blue velvety lupin, *Lupinus alopecuriodes*, is a beautiful sight with its tall spike of large flowers rising from a tasselled rosette, and its long narrow leaves tumbling all round its stocky trunk. There may also be a grassy sward, in some places thick and in others thin and broken, with tufts of species of the genus *Festuca*, a Gramineae.

Savannas and steppes

Tropical savannas usually begin at the foot of the great mountain ranges and at the borders of the larger tropical forests, particularly when these are situated far from the Equator. The soil here is dry, but not excessively so, for a few rivers and streams continue their course down from the higher regions to irrigate the savannas. These are grasslands; some consisting of grasses only, while others are dotted with trees or more frequently shrubs. Both trees and bushes have a profusion of branches but only have small, sparse leaves. More rarely groups of palms are found.

Phytogeography has shown that the continent with the largest area of savanna land is Africa. Savannas are to be found over its whole breadth, those of the Sudan being perhaps the best examples. Savannas are typical of the uplands of the Great Lake regions, and there are also great savannas south of the Equator, although they differ in appearance from those in the northern half of Africa.

The savannas that cover vast areas of Australian territory (especially in the northern and eastern regions) vary in appearance; they are often interspersed with dense, bushy, thorny scrubland or they co-exist with the sunny forests of enormous eucalyptus trees.

Savannas are not found as much in the tropical regions of Malaysia and the islands of the Indian archipelago, for here they have gradually been transformed by man into agricultural lands. Nor are they a major feature of tropical America, although there are a few in the southern part of the New World, where the climate is still hot. They are called 'llanos' (literally 'plains') along the Orinoco basin and 'campos cerrados' in the mid-west and south of Brazil.

A typical inhabitant of the African savanna is the large baobab tree (*Adansonia digitata* of the Bombacaceae family), whose sturdy trunk grows to a circumference of several yards, over the years. One reason for the thickness of the trunk is its mode of adaptation to the dry climate: its structure forms a reservoir for water, which is sparingly used by the leaves, even though the tree produces a relatively poor crop of foliage compared with the gigantic size of the branches that bear them. It is also very resistant to fire, which in the drier seasons is a continuous hazard in these areas.

Not surprisingly, the tall, homogeneous herbaceous stratum which features the baobab, the umbrella-like thorny acacias, and the thickets, also has a high proportion of Gramineae. The abundance of these woody plants often transforms the savanna into true scrubland, as in the Somali Republic, Ethiopia and – notably – in Australia, although it never achieves the dignity of a forest. More often, the savanna becomes what might be called an 'orchard' in appearance. It is fairly tidy, with tall, thick grasses growing in rows or in clumps, and small trees with wide-spreading foliage. Usually these species are from the genera *Acacia* and *Zizyphus*, and are not unlike the apple trees of temperate climates in their growth and formation.

Woody savannas are found in tropical regions where there is a winter season. Since these areas are some distance from the Equator, they are more subject to seasonal weather variations, and, as the period in which woody vegetation can flourish is so brief, there is no time for large trees with beautiful green leaves to grow on the vast plains. Even the baobabs, the elephants of the vegetable kingdom, are far from verdant, despite the water storage facilities with which they have been supplied by Nature. As has been mentioned, it is the trunk that grows to enormous size rather than the leaves, and there are many other species of so-called 'bottle trees' with swollen bellies to their trunks. Examples are *Charizia ventricosa* from Brazil and *Brachychiton rupestris* from Australia.

Savanna trees, therefore, are usually mis-

shapen, with many dry and thorny twisted branches and a few small leaves that fall shortly after the end of the rainy season, once the burning sun beats down again and parches the soil. Bushy and thicket species are to be found in greater profusion: these too are thorny and their leaves provide meagre food for the herbivores when the prolonged period of drought has dried the grass to stubble.

Succulent plants of the African Euphorbiaceae and American Cactaceae are frequently interspersed among the network of thorny bushes that lie very close to the ground. Both of these families are woody and their thick tissues are rich with mucilaginous cells that can retain water during the long dry seasons. Their stems are covered with a thick epidermis, often coated with a waxy material, to reduce transpiration and to provide protection against the vagaries of the climate. The thorns, which are usually present where there is no thick, hard bark, serve to protect the plants from the animals' unceasing search for food.

The herbaceous plants, whose fate is to dry up in the long periods of drought, owe their survival to their hypogeal organs, the rhizomes and tubers or fleshy roots which lie dormant under the soil, ready to recommence vegetative activity when more favourable conditions are restored.

Although there is so little tree coverage, the savanna – especially the orchard type – furnishes man with important materials: gums such as gum arabic which is obtained from *Acacia senegal*, resins, and oils from the balsam family from which the local people derive perfume (for instance frankincense, from *Boswellia carteri*), and dyes and drugs.

Savannas vary greatly in appearance from climate to climate and from region to region. Adjacent areas may have very different vegetation because of differences in the soil composition. The bushy savannas are of most interest to the phytogeographer, since they display the greatest variety of plants.

In eastern Africa, including Ethiopia, the 'candelabra' type of euphorbias (*Euphorbia*

candelabra, E. menelikii and others) are characteristic of the landscape, as are species of the genera *Acacia, Boswellia,* and *Commiphora,* and the baobabs and palms (*Borassus, Hyphaene,* etc) that grow up from a herbaceous grass-like vegetation consisting mainly of species from the genera *Andropogon, Setaria, Panicum* and *Tricholaena*. In the southern parts of Africa (Natal and South Africa), it is less common to find trees of any kind in the savannas, although thorny acacias manage to survive there, especially *Acacia horrida*, with its fish-shaped foliage that opens out like an umbrella.

Certain species of *Celastrus* and the spiny genera *Benthamia* and *Zizyphus* (Rhamnaceae) also grow in these regions.

In the brief period ushered in by the rainy season the grasses are tall and lush and provide excellent grazing grounds. Here again, the Gramineae predominate, in particular the genera *Aristida, Andropogon* and *Themedia,* and the scene is often brightened by flowers of bulbous plants such as the *Scilla* (squill), *Moraea* (butterfly iris), and *Gladiolus,* as well as certain Compositae, Labiatae and Polygalaceae.

The Indian savannas also contain Gramineae, although the genera *Saccharum* and *Imperata* predominate. A few gigantic teaks are to be found there (*Tectona hamiltonii* and *T. grandis*) as well as specimens of woody Verbenaceae. Acacias are common (especially the *Acacia catechu*), as are the *Albizzia, Dalbergia* and *Bauhinia,* all Leguminosae, and the *Terminalia* (Combretaceae), whose bark is used for dyes and tannins. Palms, bamboos and Cyperaceae also grow in the less arid areas.

Lastly, one should not overlook the savannas of South America that cover most of the enormous territory extending from the 'catingas' of northern Brazil to the Brazilian and Argentine 'chaco', almost to the edge of the pampas. There, thick grassy vegetation quite often gives way to the Bromeliaceae of the genera *Bromelia* and *Achmea* that thrive in a hot, dry climate. Among them grow the thorny, dwarf *Cereus* (torch thistle), in some places so dense and thick as to

*remarkable, spire-
shaped termites' nest
forms a landmark in the
tropical savanna
country of central Africa*

form impenetrable barriers to anyone who attempts to pass through them. There is no shortage of spiny plants; indeed, they are so abundant as to give this area the name 'espinar'. As always, they contain umbrella-type Leguminosae (*Acacia, Cassia, Mimosa, Caesalpinia, Prosopis*), and species from the genera *Spondias* (hog plum), *Zizyphus, Capparis* (caper bush), succulent euphorbias and a few palms of the *Mauritia* genus. The Bombacaceae are represented not by the baobab but by other trees with water storage trunks such as the *Cavanillesia*.

By comparison with the vegetation of the savannas, that of the steppes is much poorer both in variety and in quantity. There are steppes in every latitude and at all altitudes: in Arabia, Anatolia, and Tibet. The term has a very specific meaning, referring to the vast expanses of land, mainly plains covered with herbaceous, usually graminaceous, vegetation, especially species from the genera *Stipa, Calamagrostis, Poa, Spartina*, and often including Leguminosae (*Trifolium, Medicago, Astragalus, Melilotus* etc), and Compositae (*Artemisia, Achillea*).

The herbaceous vegetation of the steppes is often bushy, partly annual and partly perennial, and the various species have rigid though slender stems whose tissues consist of hard cells. This is the reason why the vegetation is also called 'duriherbae', or 'hard grass'. The steppes are enormous, fertile terrains. They provide pasture for a great number of herbivorous animals native to the different regions, from cows to goats, and from sheep to camels.

Although it is so monotonous, there is a wide variety of vegetation in the steppes. Where there are trees, however, the area is similar in physiognomy to the savannas; in other places the steppes resemble prairies, lush with herbaceous plants with large succulent leaves. Finally, the steppes may contain very sparse vegetation with clumps of thickets in areas that are otherwise almost bare.

Not a bush, but another large termites' nest in woodland in Queensland, Australia

Nonetheless, the steppe is normally associated with certain environmental factors such as poverty and dryness of soil and very high daytime temperatures, apart from seasonal variations, in the steppes in temperate or cold regions. The air tends to be very dry and the wind blows almost unceasingly.

In steppe areas, the quantity of nutritional salts may be equivalent to the salts in good meadow land, or only slightly lower, but the lack of water means that they cannot be absorbed by the plants. Very often the structure of the soil and its surface causes rainwater or any running water to disappear immediately, either sinking into the ground or evaporating. In consequence, substantial quantities of salt residues accumulate, making the earth extremely saline. Obviously it is impossible for most herbaceous plants to survive – especially when the salt in question is sodium chloride – and the only ones that do are the plants which botanists describe as halophile or salt-loving. Moreover, when sodium chloride solutions react with calcium carbonate in the soil, sodium carbonate is formed and the soil becomes highly alkaline. Few plants can adapt to this.

In these terrains, the vegetation is characterized by species of the genera *Statice* or sea lavender (Plumbaginaceae), *Salicornia, Suaeda* (Chenopodiaceae), *Glaucium* or sea poppy (Papaveraceae), *Plantago maritima* and certain more specialised members of the Leguminosae, Umbelliferae, Asteraceae and Gramineae families.

These in brief, are the features of steppes in general. Most factors are more extreme in tropical regions because they are intensified by the high temperature which causes a considerable loss of moisture, but there are few such tropical steppes. They are limited to certain sectors of Central America (Mexico) and South America (Colombia, north Brazil), and southern zones of Central Africa (Angola, South Africa), as well as the coasts of Arabia (Yemen, Oman), Sinai, the more southerly regions of Iran and extensive areas in Australia (notably Queensland).

Plants of the desert

The most extreme type of environment from the point of view of land vegetation is the desert. The word evokes an image of vast expanses of terrain without trees, shade, or water, where rain never falls and the wind never ceases to blow, wearing away any rocks that exist by its tireless action, or shifting enormous masses of sand to form dunes. The sand, as it moves, uncovers and buries again the remains of animals or the few withered prostrate bushes that manage to scrape a meagre living.

Deserts exist in all latitudes and altitudes. High-altitude deserts, which are caused by extreme cold, often have a permanent covering of snow or ice. There are the frozen wastes of Greenland, the Antarctic and the high mountain ranges which never shed their snowy mantle; the Gobi desert in Mongolia, for instance, is situated at approximately 45° North. These so-called cold deserts are a completely different phenomenon from the desert regions of warm areas.

There are tropical deserts in north Africa (Sahara) and South Africa (Kalahari), in the central part of Arabia, in Arizona and Utah in North America, and in Australia – the Great Sandy Desert and the Great Victoria Desert. In fact, Europe is the only continent without any desert region, although there are some arid areas around the Black and Caspian Seas and in parts of the Ukraine and the North Caucasus.

Climatically, deserts are essentially a feature of continental interiors, even though some stretch almost to the sea, as does the Sahara, which extends to the Mediterranean in the north and to the Atlantic in the west. Of all environments, deserts have the lowest rainfall, water is short and vegetation is reduced to an absolute minimum or even eliminated altogether, as in the middle of the sandy Sahara.

Tropical deserts which include mountain areas are called rocky deserts; some have mountain peaks, others have monstrous glaciers and broken cliffs. Even when streams descend to the foot of these mountains, they are small and the water they bring rapidly disappears as it reaches the plains because of intensive evaporation and the inability of the sandy or rocky ground to retain the water, which filters down and is dispersed.

Deserts and sub-deserts occupy 8.77 per cent of all the total area of land, a substantial proportion of the Earth's surface, and are said to represent an extreme condition, or 'ultimate stage', the preceding stages being the forest formations such as the 'espinar' or the prairie configurations of the steppe and savanna.

Rain is extremely rare and is not seasonal; the periods of drought often last more than a year. Taking the Sahara as an example, the temperature ranges from 52°C (about 125°F) in the shade to 70°C (about 160°F) in the sun during a summer's day, dropping to 10°C (50°F) during the night. The wind blows continually, eroding the rocks and increasing the high degree of evaporation so that it is barely possible for vegetation or animals to survive.

The only resources that could alleviate such extreme dryness are dew, however small in quantity, and underground water. Where the latter comes to the surface, oases spring up like green jewels in the formless expanses of sun-baked land.

The vegetation at oases can play an extremely important role in fostering more plant life, since dew forms where there are plants, and vegetation produces vegetable waste (and animal waste, since animals live there). These will in turn form a layer of organic matter useful not only for the establishment of pioneer microflora and microfauna but also for the absorption and retention of water, whatever its origin.

Above right: A desert in the south-eastern regions of Egypt. The vegetation is sparse and is to be found only on isolated sandy mounds. Below right: Semi-desert dotted with acacia in the Agordat region of Ethiopia

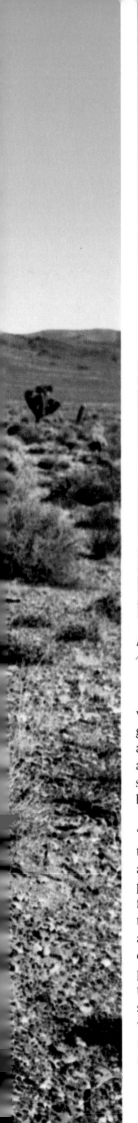

Left: Scrub desert in Death Valley (USA).
Above: A natural hedge of tall Cereus

Although the type of desert vegetation may vary from continent to continent, there is one great and undeniable common factor: it is poor and sparse. There are very few plants, their organs are smaller than normal, and they are widely spaced out in large areas of bare soil, which may be sandy, icy, or rocky.

Deserts have been described as the extreme 'paroxysmal' forms of the original environment that have evolved by very slow deterioration, and undoubtedly vegetation is involved in this process. Plant life has been gradually transformed until it has disappeared altogether, contributing to the worsening of climatic conditions and finally creating absolute desert. That the deserts were once covered with vegetation is proved by the petrified remains of large tree trunks that sometimes raise their heads above the sand as it is shifted by the wind. Some deserts are very salty, and are comparable to the saline steppes. They contain substantial quantities of sodium chloride, which often react with the calcium salts to produce sodium carbonate and eliminate plant life. One example is the Natron desert in Egypt, where the salt of the same name (natron, known to the ancient peoples) was collected.

Excessive salinity is generally due to the fact that when rain washes the mountains around the desert zones or the masses that form the rocky deserts, large quantities of salt solution run down to the lower level together with the detritus, producing deposits that crystallize, because of the rapid evaporation of the water, and give rise to surface efflorescence.

Thus, the major factors that conspire against desert life are the interaction between the soil and the environment, the very high temperature, the effects of wind, and the permeability of the overlying atmosphere to light rays (including ultraviolet rays). In these circumstances, very few plants can survive.

A few, however, do manage and grow to a reasonable size, bearing small green leaves. One such plant is the tamarisk, which is found in places in the Somali desert, although it is never far from water, either on the surface or under-

Right: Succulent plants in flower: Rebutia senilis, a cactus from northern Argentina (top), and Euphorbia hermentiana (bottom)

Left: In the desert spurs of the Andes, dry shrubs and thorn are interspersed with tall species of Cereus

Left: Round leafless desert cactus stalks bristling with spines. Two of the numerous Echinocactus species are shown here

ground. Indeed, the presence of water can be detected by the arrangement of the tamarisk shrubs, sometimes in fairly compact groups and sometimes strung out along what might be called an imaginary underground stream. In other cases, the desert suddenly explodes with flowers, often borne by strange plants.

In African deserts, there is an abundance of aloe with beautiful rich plume-like inflorescences, tubular yellow, orange, or red flowers living in association with the spiny cactiform euphorbias or large dragon trees (*Draceana draco*). Species of the genus *Stapelia*, or carrion flower, of the Asclepiadaceae family, live in India and in the few desert lands of Madagascar; these too are cactiform, with strange blooms which give off a fetid odour.

One of the most remarkable inhabitants of the desert is the *Welwitschia mirabilis* (*Tumboa bainesii*) of the Welwitschiaceae. This plant is a 'living fossil', and the sole surviving species of the family. It inhabits the desert and subdeserts of tropical west Africa and southern Angola, in an area that extends over a strip of land approximately 600 miles long and 125 miles wide. This plant, which may live for more than a hundred years, has a long, sturdy tap root that is joined at surface level to a short, tuber-like stem which grows to a diameter of more than three feet (one metre). Two long, wide ribbon-like leaves grow from the stem, one on either side, and creep outwards along the soil. These leaves, which have parallel venation, are leathery and thick but their tips wear out and become frayed as they grow longer on the ground. The plant is unisexual and the flowers, cones about a foot high, are borne on stalked, ovoid spikes. *Welwitschia* is in danger of extinction and is now protected by stringent conservation measures.

Almost as strange are the plants which combine together, interlacing their small trunks and branches to form cushions of growth; they have thorns but few leaves, as a result of environmental conditions, although they may produce many flowers, as in the case of the plant *Arabis aretioides*.

75

Left: A cactus from the American deserts, Opuntia microdasys, from New Mexico. Bottom: Two more cacti: An Opuntia in flower (left), and Brasilocactus haselbergii (right)

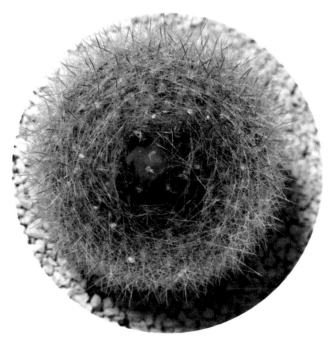

It is not uncommon for a barren desert to become covered with sparse vegetation in the period immediately following the brief rains that fall without warning. These so-called 'ephemeral' plants germinate and flower, and their seeds ripen within a few days. Although their life is so short, it is sufficient to ensure survival of the species, which is probably in less danger of extinction than less remote plants in more populated areas.

The northern plateau region of Mexico has slightly less extreme desert conditions, as do the adjoining areas of Arizona, New Mexico and Texas and here there is a flourishing growth of mesquite, greasewood, creosote bush, yucca, and various species of cactus.

Survival at all costs

It is interesting to observe the curious ways in which Nature enables plants to adapt to adverse environmental conditions such as those of the desert. Some plants, for example, have a very limited vegetative apparatus above ground, while their root apparatus is unusually well-developed and ramified, digging deep into the soil to absorb a substantial quantity of water, when it is available for very short periods only on rare occasions.

Also, as in the case of the steppe plants, desert vegetation can reduce transpiration to the minimum by various modifications either in the formation of small leathery leaves, whose surface is sometimes coated with a thick cuticle and protective down, or by producing thorns.

The stems of desert plants may become water reservoirs of columnar and swollen appearance, or they may have ribs or wings down the sides; they are often covered with hairy bristles, wool, or spines. This feature is exaggerated in what are commonly known as the succulent plants, the Crassulaceae, Cactaceae, and Euphorbiaceae. Because of their cell structure, they possess the ability to accumulate large quantities of water

Water determines the existence of plant life in the deserts. Where it flows, plants abound and Man settles. Right: The large spring that serves the oasis of Ghadames in Libya. Below: 'True' desert vegetation: a sea of sand and pebbles stretching as far as the eye can see, in which thorn struggles to survive

in moisture-retaining tissues, so that in some cases they can tolerate years of complete drought.

The most remarkable plants in this respect are some of the many Cactaceae native to the deserts of tropical America – cacti that are now grown all over the world as ornamental plants, because of their strange shapes and beautiful flowers.

Only in rare cases, however, usually under glass in botanical gardens, do these plants attain such sizes as in the places of their origin. Certain specimens of *Cereus*, for instance, may achieve monumental proportions in their own habitat.

There have been examples of *Cereus giganteus* that were no less than 15 metres (50 feet) in height, which could store up to 500 gallons of water. There is also one recorded case of an *Echinocactus* from Arizona which, when deprived of the water that it had accumulated, lost only 13 per cent of its weight during a research project lasting many months.

Succulent plants have been known to students of botany since the sixteenth century, when the first rare specimens were imported from distant lands. At first, these were cultivated only in convent gardens or by pharmacists, but later became favourites of lovers of tropical flora.

The Cactaceae provide amazing examples of the ways in which plant organs adapt to extreme biological and ecological conditions. It is inter-

esting to consider the methods of adaptation in the light of the truly incredible forms that such plants may assume in their natural habitat.

The list of 'living phenomena' should begin with two species from the genus *Astrophytum*. The first, *A. myriostigma* (more commonly known by the name of bane berry), is a native of Northern Mexico. Although it is almost spherical initially, it grows into a shape like a bishop's mitre and has a maximum height of 50–60 centimetres (about two feet). The plant body has rough shaped ribs; it is greenish gray, and spineless, but is covered with minute tufts of white down; the flowers are fairly small, with silky petals, blossoming only on the apex.

The second species, *Astrophytum coahuilans*, is perhaps only a variety of the former, but its area of distribution seems to be limited to the Mexican province of Coahuila; it is not very different from *A. myriostigma*, except that it is shorter, more scaly, and the flower is red inside.

A member of the same genus is *A. asterias*. Although its name indicates that it is similar to a starfish, it is in fact more like a sea urchin. The body of the plant resembles a flattened hemisphere and consists of eight to ten segments, each having a line of wool-covered warts down the slightly recessed middle line. During the long periods of drought by which its habitat is often

ravaged, this cactus contracts and shrinks down into the soil, becoming almost invisible amongst the pebbles. When it blossoms, on the other hand, it produces on its apex a large lemon-yellow flower which consists of several lanceolate petals that are red in their central portions.

No less strange is *Ariocarpus trigonus*, also restricted to certain regions of Mexico (Coahuila, Tamaulipas, Nuevo Leon). When fully grown, it is like a short grayish tree trunk with irregular scars all around; its top portion breaks out in numerous large greenish-gray protuberances like splinters of stone, some pointed and some more rounded, two to five centimetres long (up to two inches); in the middle of the rosette formed by these strange protuberances lies a cushion or thick tuft of whitish wool from which the fairly large yellow flowers emerge. All *Ariocarpus* are poisonous, but are still used in Mexico in

the preparation of certain medicinal remedies.

The plants *Strombocarpus* and *Encephalocarpus* are also short-bodied cacti, and each of these genera has one species of great interest. *Strombocarpus* (*Echinocactus*) *disciformis* from Hidalgo, Mexico, looks like a large greenish-gray fish; the ribs only just project and spiral upwards, and form a close-set series of solid, fleshy, diamond shapes in the centre of which are pimple-like areoles. The areoles in the apical portion have long, rigid, erect bristles; they are grayish in colour and short-lived. This is the region of the plant in which the yellowish-white flowers blossom. They are made up of lanceolate petals that are yellow at the base and red at the tip.

The *Encephalocarpus strobiliformis* is also a Mexican species, being found in Miquihuana and Tamaulipas. It has a more cylindrical, erect

Two examples of 'protective' adaptation in desert plants. Above left: Espostoa melanostele, a cactus with a protective 'hood' Above right: The flower and large succulent leaves of a Mesembryanthemum, from the extreme south of Africa

Left: Two further examples of plant adaptation to desert life: a leafless thorn bush and a plant whose small leaves, rather like heather, prevent excessive loss of moisture

body consisting of a lower, almost club-shaped, woody gray portion and a globular greenish upper portion. Both have a spiral series of scales which are more marked in the upper than in the lower part. Each scale is crowned by a small areole out of which grows slender spines. The flowers of this species are a very gaudy carmine colour.

There is insufficient space here to dwell on the numerous species of *Mamillaria*, which all have globe-shaped or cylindrical globe-shaped bodies.

Some, like *M. woodsii*, bear small red flowers arranged in a circle, while others have sparse, large, pink flowers, hooked spines and tufts of soft white bristles (for example *M. guelzowiana*), or are entirely covered with feathery white spines so that they look like balls of fleecy wool, (*M. plumosa*). Less woolly and without the 'halo' effect of others is *M. bombycina*; its clusters of plant bodies, its white, glossy spines, and its numerous cyclamen-coloured flowers make this one of the most beautiful species.

Another interesting cactus is the *Dolichothele* (*Mamillaria*) *longimamma*, also a native of Mexico, whose greenish body consists of many long cylindrical, angular, fleshy tubercles five to eight centimetres (two to three inches) long, each terminating in a long, central, needle-like spine surrounded by more slender spines in a star shape. The flowers are large and yellow, each consisting of lanceolate-tipped petals.

There are many other strange and wonderful cacti, including the *Zygocactus*, *Epiphyllus*, and *Phyllocactus* with their branching stems, and flat segments like smoothy, fleshy leaves with large flowers at the extreme tip, *Cephalocereus*, the irregular *Neoporteria*, and *Medlolobivia*, and the extraordinary *Opuntia clavarioides*, whose segments have lobes shaped like the fingers of a hand.

To return to plant adaptation: another way in which Nature meets the need for the perpetuation of desert species is by ensuring survival when the life cycle is terminated or interrupted by the prohibitive conditions. The organs of reproduction are vital, and they can survive because it is possible for them to be uprooted and borne away by the wind. This is the case with *Asteriscus pygmaeus*, a minute asteracea only a few centimetres tall at the most, which has a relatively long root but few or no branches. It has a very few leaflets arranged in rosette form and ten or more flowers producing rapidly maturing fruits at the tip of short twisted twigs. At first they retain the seeds which emerge from the theca only after they have been blown from the plant by the wind and carried great distances.

The plants of a member of the Cruciferae, *Anastatica hierochuntica*, known as the Rose of Jericho or the resurrection plant, are also windborne, as are selaginellas found in the deserts. When the dry period lasts for a long time, the branches of these small plants form a close-set ball about as big as a fist. As the branches bend inwards, they protect the spore-bearing organs of the *Selaginella*, or the gemmae in the case of *Anastatica*. Curled up like a hedgehog, these curious 'balls' can be made to roll and jump by

the wind over long distances in the desert. They
have two chances of survival: either the plants
may start to grow again when they come to rest
in moist or muddy terrain, or if the ball itself is
not able to take root in the soil, then the spores
can be spread around. This phenomenon will be
assisted, or rather forced, by the fact that the
branches open out whenever the night air is cold
or damp, closing up again next morning. The
sequence of movements is evidently due to the
hygroscopic nature of the branch tissues.

Other plants, including the tamarisk, secrete
droplets of calcium chloride or magnesium
solution through the skin of their leaves which
are very hygroscopic. The purpose of these
secretions is to attract and retain humidity from
the air. This moisture, though slight, is the product of condensation resulting from a drop in
night temperature. The plant can then use the
water it has absorbed as and when it needs it.

With other plants, especially the Gramineae, the hairs are not only a means of protect-

ing them against excessive transpiration, but also
provide a felt-like layer that can retain humidity.

The plants that live in the dunes of the sandy
deserts have to be even tougher, for they have a
harder struggle to survive, since the substratum is
not only extremely dry but is also constantly
shifting. During infrequent rain, the water
filters through the grains of sand immediately,
and if any seeds that happen to fall there manage
to germinate they first develop a disproportionately long root apparatus. Sometimes, in
species of the genus *Aristida* (Gramineae), for
example, which adapt to this life, the roots with
their dense, extraordinarily ramified branches
extend over a radius of approximately 65 feet
from the parent plants. This emphasizes two
features: the ability of these plants to seek out all
the water available over a vast radius, and the
efficient way in which they manage to grow in the
sand of the dunes. Many Gramineae that live
in a sandy habitat are in fact grown on purpose
to stabilize the dunes of deserts and sea coasts.

Strange alliances

The curiosities of Nature are to be found in greater abundance in the tropics than in the plant communities of temperate or cold areas, and nothing more strange than the 'allegiances' that may form between some species of tropical plant and a particular animal, for quite commonly plants and animals congregate and cohabit in involuntary societies, called 'symbiosis' by the naturalists.

A symbiotic union is quite different from a parasitic relationship, in which the host may actually suffer damage as a consequence of the presence of the parasite. It is rather a 'mutual aid association', in which the two live together to their mutual benefit. There are many examples of this, to be found in all forms of life, from the small birds that live on the back of the rhinoceros, to the relationship between the hermit crab and the sea anemone, and the fig tree and the fig wasp. In all of these cases, both of the 'partners' derive some benefit from the union, although they are capable of living totally separate existences.

One of the most outstanding cases of all, however, is the cohabitation of certain ants and *Acacia cornigera*, one of the many species found in savannas and steppes. The thorns at the base of the leaves, like stipules, are swollen at the point of insertion; despite their limited surface area, all the thorns contain two small thin-walled sectors. The ants are aware of this structural peculiarity and because the tissues are easier to break down, being more tender and flaccid, they excavate short tunnels to provide access to the nest that they build within the base of the large pairs of thorns.

Immediately above the thorns, at the point where they join the leaf petiole, there is a minute oval trough, in the bottom of which is secreted a thick sugary liquid or nectar. The troughs in fact are called 'extra-floral nectaries' to distinguish them from the 'floral nectaries' found in the flowers themselves. Naturally the petiole also forms the rachis of the bipinnate acacia leaf composed of many small leaflets. At the tip of each of the latter is a small globular projection known as 'Muller's mouth', after the naturalist by whom it was discovered and interpreted; as the ants explore the twigs and leaves, they remove these tiny, tender appendages and feed on them without damaging the plant in the process.

This association between ant and acacia is not the only form of co-operation between insect and plant. Ants are true warriors by nature, and they live permanently on the tree which they 'patrol', defending the few small leaves from the herbivorous animals which try to eat them, either because they choose it or because it is the best they can find. They also protect the tree against the leaf-cutting ants.

The alliance of other plants with ants of different species is also common. There is, for example, a genus of tropical plant (*Myrmecodia*, Rubiaceae) and others, some of which live epiphytically on the branches of large trees, while others are terrestrial. They have a stem which is rendered even grosser in shape by the ants that live within, and which swell the internal organs of the plant by their nests, which are bored into the heart of the stem, sometimes to quite a considerable depth.

When an animal or man comes into contact with the Myrmecodia, these insects demonstrate their well-known collective aggressiveness by emerging from their hiding-places in hundreds, falling on the unsuspecting invader and torturing him with their bites.

Other examples of myrmecophily, or the friendship between plants and ants, are provided by species in the genus *Cecropia* from the forests of tropical America. In somewhat the same way as the acacias, these plants have branches with scar-shaped nodules which indicate to the ants

the points at which they can bore their tunnels and build their nests. The fierce ants install themselves there and defend the plant; at the base of the leaf, the plant has a tuft of soft hairs prized by the ants as food. A similar case is that of *Tococa gujanensis*, from the American tropics.

The fearsome thorns of one of the many acacia species forms an effective defensive barrier for the tree

Plants that feel and eat

Whether the capacity for feeling is possessed only by the animal kingdom is a concept open to discussion; without doubt, 'feeling', in the sense of awareness of what is happening within and without a living organism and the reactions aroused in response to stimuli, is a phenomenon peculiar to the animal world, or rather to advanced animals. But if the word 'feeling' means all those properties that the living cell – both animal and vegetable – possesses, enabling it to receive impressions and stimuli from the outside environment and to react, then the phenomenon applies also to plant life. In some cases, plants react so strongly to stimuli that the responses are immediately recognizable. Here again, the tropical regions provide a greater wealth of examples than do the regions that have more temperate climates.

It is, of course, essential for any living thing to react to its environment and interact with it; indeed, the capacity to do so has been the basis of one definition of life itself. In order to respond to the environment (and this applies to plants as much as to animals), the organism needs first a means of detecting, or sensing, stimuli; secondly an internal communication system, so that feeling can be translated into action; and finally, the apparatus that produces the response. In animals, these are, respectively, sense organs, the nervous system, and motor or effector organs. In plants, on the other hand, although the system is fundamentally the same, the sensory and effector organs may be one and the same, and where there is a need for communication it is achieved very simply.

The basic needs of life which necessitate these 'feeling and doing' systems are numerous: reproduction, respiration, etc. But in the case of plants, the two that sometimes elicit a visible and active response are defence and nutrition. We will consider examples of both: the *Mimosa*, which draws back when 'threatened' by touch; and some of the species which trap insects for food and which are so surprising and spectacular in their cunning.

Mimosa pudica or sensitive plant is one of the best examples of plant sensitivity. Its scientific name is indicative of this quality, for it is derived from the Greek *mimos*, meaning mimic. It is a small, bush-shaped tree belonging to the Leguminosae family, and is a native of Brazil, from where it was imported into Europe in 1638. Related species (*Mimosa spegazzinii* and others) also have this property of irritability. These mimosas have compound, bipinnate leaves consisting of many leaflets. When they are moved by some object or even when they are merely shaken by the wind or stimulated by a small flame the leaflets on either side fold back against each other and draw closer to the rachis, taking up their night-time position.

The explanation for this strange reaction to stimulation is thought to be that the leaves of the mimosa have special tactile hairs inserted above cushions of thin-walled cells. These hairs are capable of transmitting the impulses from the stimuli to special cells located either at the base of each leaflet of the 'motor pads', or pulvinus. These cells contract and cause the leaflets to move. It is surprising, however, that if only one of the hairs on the pulvinus is stimulated the leaflet in question moves, and if a single leaflet is touched all the neighbouring leaflets will fold back. Rougher treatment naturally causes the whole leaflet to bend backwards, and if the impact is even more violent all the leaves on the branch, or even the whole plant, will fold back.

In the light of present-day knowledge, it is interesting to return to the views of the naturalists of the past. They considered the sensitivity of the mimosa to be somewhat similar to the process

Sarracenia purpurea, a curious insect-eating plant from California

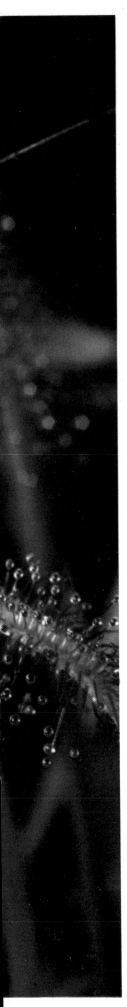

occurring in animals having a nervous system. More recent biological and physiological research has shown that remote transmission of stimuli in the *Mimosa pudica* takes place as a result of a special substance, as is demonstrated by a very simple experiment. If a whole sensitive plant leaf is cut off at its base and placed in water, it can be seen that the leaflets that have folded back as a result of the shock of amputation will gradually return to their normal position. If without causing the slightest ripple a little juice from crushed mimosa leaves is added to the water, a few minutes later the leaflets start to fold back as if they had been stimulated by a slight impact. The leaf with its petiole in the water merely has to absorb a little juice of the 'ill-treated' leaf organ to provoke a reaction in the pulvinus. If a stem or twig of *Mimosa spegazzinii,* equally as sensitive as the other species, is cut at a given height and the two ends are joined by a tube made of glass or other inert material filled with pure water, the plant tissues of the two pieces will obviously connect artificially through the medium of the water. If the lower part of the mimosa is then stimulated, after a little time the leaves situated above the break will start to react, demonstrating once again that it must be a liquid substance that is charged by irritation and that by this means the irritation is transmitted to the pulvinus.

There are many other examples of the way in which advanced plant organs react to stimuli, the stamens of the flowers of certain genera belonging to the Berberidaceae (notably *Berberis* and *Mahonia*), for instance. However, the sensitivity of many plants defined as 'carnivorous' (which should more accurately be termed 'insectivorous'), is of even greater interest. The main species belong to three individual genera, *Dionaea* (Venus fly-trap), *Drosera* (sundew) and *Pinguicula* (butterwort, bog violet). They are not all tropical nor even all exotic, since certain specimens grow in temperate regions. As in the

A drosera from the Cape of Good Hope (Drosera capensis). Sensitive red hairs, like pins, cover the leaves

case of the sensitive plant, the movements of the species in these three genera are reversible, in that once the stimulus ceases their sensitive organs return to normal.

Dionaea muscipula (Droseraceae), for example, is a herbaceous plant of North America, particularly Carolina. Its peculiar ability to catch small insects in its sensitive leaves was recorded by Linnaeus by the words '*miraculum naturae*'. This herbaceous plant lives in damp, mossy places on the banks of stagnant water. It is about eight inches (20 centimetres) high and has a basal rosette consisting of a few strangely shaped leaves, pressed close to the ground. From the middle of the rosette rises a stem with a clustering inflorescence of five to ten white flowers. Its leaves, however, are its special feature, as each has a basal blade like a wide petiole, and a larger apex which opens into two almost semi-circular symmetrical lobes with a fringe of rigid appendages like bristles, some coloured green, some a beautiful crimson. In the middle of each half of the leaf are three stiff, long hairs, the 'trigger hairs', which are attached to a mobile pad similar to the pulvinus of the sensitive plant. Tiny purple or glossy glands secrete a tempting viscous liquid on to the lamina, while star-shaped hairs at the edge complete the trap for the unsuspecting victim.

Insects, attracted by the shiny leaves and their gay, contrasting colours, try to alight on the upper portion of the leaf which closes like the pages of a book at the slightest contact. The more clumsily an insect touches any one of the six bristles on the lamina, the quicker is the leaf's reaction: often only 15 to 30 seconds elapse before the trap closes and the teeth in comb formation on one side of the leaf lock into the spaces on the opposite side.

The most amazing thing is that if wind or some other outside factor happens to drive a chance piece of flotsam into the leaf – for example a dry twig, a grain of sand, a scrap of paper, or any particle without nitrogenous food substances in it – the leaf will not stay closed on it, but will reopen within a day or two. If, on the other hand,

it traps an insect, whose body is full of proteins, the leaf edges clamp together even more closely, as if to squash the insect, while at the same time its microscopic glands secrete a viscous acid mixture, rich in proteolytic enzymes which cause the breakdown of proteins into simpler substances. This enzyme mixture dissolves the softer parts of the victim, and the resulting solution is slowly absorbed by the plant through special leaf cells. The period taken for this extraordinary digestive process varies from about ten to thirty-five days. After this time, the leaf slowly opens up again, the hairs stand erect, and on the leaves the chitinous remains of the insect can be seen. These are dried by the sun and air, and sooner or later they are blown away by a gust of wind, leaving the leaf clean and ready for the plant's next 'meal'.

One of the genus *Drosera, Drosera rotundifolia,* is found widely distributed in temperate regions amid clumps of sphagnum, and many species of this genus live in similar habitats in different regions of the world. *Drosera intermedia* has a predilection for temperate regions and is rather rare, while *D. longifolia* is an Alpine plant; some species are found only in the Australian continent, while others are natives of India, Madagascar, or North Africa.

The leaves of most of the plants of the genus *Drosera* tend to be arranged in a fairly regular rosette, which is situated at about ground level, where it is cool and mossy; this rosette measures from five to eight centimetres (two to three inches) in diameter at the most, and from its centre rise one or more (though never many) slender scapes bearing small white blossoms. Exceptions to this pattern of growth formation are found, however, in three species. The leaf of *Drosera rotundifolia* has a long petiole and a concave disc-like lamina, while the lamina of *D. longifolia,* as the name suggests, is elongated and narrow, and *D. aliciae* possess a spatula-shaped leaf which widens at the apex. With the *Drosera,* the leaves are the entrapping agent, but unlike the *Dionea* they have very many sensitive hairs. The leaflets – whether round and slightly concave or elongated as in *Drosera longifolia* – have long petioles, on the upper side of which lie

Above: A sprig of sensitive plant (Mimosa pudica). The leaves are normally open, but collapse at the slightest stimulus

many carmine hairs standing out in all directions, each one with a small swollen pin-shaped head. These 'pin-heads' are coated with a viscous liquid which makes them sparkle like drops of dew that do not evaporate even on the hottest summer day. This secretion is sweetish and is produced by the heads of the hairs themselves, which are in fact glands. The colour and glossi-

ness of the hair-covered leaves tempts the insects to come and investigate what they believe to be nectar. They adhere to the sticky liquid and all the hairs converge, curving in upon the tiny prisoner which becomes enmeshed in all these tentacles and undergoes the slow process of digestion already described in the case of the *Dionea*.

There are many other insect-eating plants; they include *Sarracenia purpurea* (the pitcher plant, side-saddle flower), *Heliamphora nutans, Darlingtonia californica* and the pitcher-leaf plant (a species of the *Nepenthes* genus), which are certainly worthy of a brief mention.

The first three species belong to the Sarraceniaceae family and are to be found in many locations in both North and South America, in marshy areas. They have a rosette of basal leaves which are rolled lengthwise to form a long conical tube, opening out slightly at the top. Inside this ascidium, or pitcher-shaped leaf, are longitudinal rows of hairs that form ridges, like folds. At the top, the lamina expands and is often folded over to form the 'lid' or operculum, of the pitcher.

In the case of the Sarraceniaceae, insects are attracted by the secretions which gloss the inside of the ascidial 'pitchers'; outside, these structures are green but attractively veined and tinged with red. The viscous liquid is retained on the ridges of hairs, making them slippery, especially as they slope downwards. It is obvious that any small insect imprudent enough to enter the well-baited trap will slip in, and will fall to the bottom into the pool of thick liquid that will then dissolve the soft parts of the victim.

It is not entirely clear how the remains of the insect are dissolved, digested and then absorbed. Some botanists believe that the insect-eating plants in this group do not produce any substance to dissolve the proteins and that only after the corpses have decomposed through normal extraneous processes (proteolytic bacteria, etc) and by rainwater, can they be absorbed, but scientific opinion on this point is not yet unanimous.

The pitchers of the Darlingtonia, an insect-eating plant from North America that recalls some strange fairy-tale monster

While the ascidial structures in *Sarracenia* are as described above, in *Heliamphora* they tend to be more cone-shaped, and in *Darlingtonia* they have a long and grotesque-looking drooping fish-tailed flag.

An even stranger insectivorous plant is the *Nepenthes* (Nepenthaceae). It lives in the boggy regions of equatorial forests in the Old World and the Sunda Islands, Madagascar, Seychelles, New Caledonia and tropical Australia. These plants, of which there are about seventy species, have long petiolated downward-pointing leaves; after the lamina, the widest and longest part of the structure, they have an appendage similar to a petiole terminating in a pitcher-shaped organ, again known as an ascidial structure. This sometimes has an operculum, or lid, and a thick liquid in the bottom of the deep urn gives out an odour that attracts the victims.

These ascidial 'containers', often in bright, mottled colours, set their trap with a large rim curved inwards around the edge. The orifice secretes a sort of nectar which entices the insect. The latter comes to rest on the edge of the container but immediately falls into the liquid in the bottom, because the edges are rounded, steep, and slippery, there to meet its fate.

In *Nepenthes* where the ascidial container has a lid, the operculum is jointed at a point of the orifice and may move up or down, depending on whether or not the plant wishes to prevent insects entering or escaping from the fatal trap.

Linnaeus apparently named these plants *Nepenthes* as an allusion to their medicinal properties, although it now appears that in fact they have none. Nor can the Egyptian plant with the same name mentioned by Homer, meaning 'painless', or 'grief-assuaging', have any connection with the deceitful insect-eater of the humid virgin forests.

Timber, fibres, and forests

The great marshy forest of the limitless Euro-Siberian plain, or 'taiga', provides an enormous store of resinous wood, since it contains firs, larches, and pines. However, the natural stores of timber which the tropical forests contain in their intricate network are no less impressive. Man has great need of such exotic woods as teak or iron wood, mahogany, rosewood, ebony, and many others.

The time is long since past when the main use of timber, especially in the cold countries, was for fires to provide warmth. Even more remote are the times in which almost the whole surface of the earth above water level was covered with dense forests. These forests still exist in tropical regions, despite the fact that man is felling these trees and replacing them with crops and towns, as he has been doing for thousands of years in the temperate regions of the Old World. Man's colonization of what used to be the wilderness of the natural environment may have brought about social and technical progress for the local peoples, but at the same time it has undoubtedly helped to destroy the natural resources constituted by the forest and its fauna.

In many cases, indeed, the discovery of precious forest substances has sparked off the process of destruction, as in the case of the forests of African mahogany on the Ivory Coast, the Gabon okoumé, the ebony and various hardwoods from Madagascar, the teak from Indo-China (*Tectona grandis*), and the northern conifers both in Eurasia and in America.

These woods form a vital group of raw materials, not only as a source of heat when used as fuel, but also for the production of charcoal, pulp for the manufacture of paper and cellulose, textile fibres, resins, colophony, balsam, essence of turpentine, rubber, oleoresin, xylene, phenol, guaiacene, creosote, and other substances for pharmaceutical and therapeutic uses, acetic acid and other industrial acids, acetone, formaldehyde, various solvents, dyes, tannins, and so on.

The word mahogany, or 'acajou' in French, is used for certain exotic timbers originating in Africa and America. The most important African genus is the Khaya mahogany of the family Meliaceae (*Khaya ivorensis, K. grandifolia, K. senegalensis, K. anthothea*) from the Ivory Coast, Senegal and from Cameroon. The warm red or golden wood is satin in appearance after it has been worked, and is highly prized both for the making of beautiful furniture and for inlay work. For some time now it has been used in the production of plywood. True mahogany, however, comes from a plant from the New World, the *Swietenia mahagoni*, which grows in the forests of Cuba, Nicaragua and Honduras. There are other related species, such as those belonging to the genus *Entandrophragma*, known locally as 'tiama', 'sipo', 'aboudikro', and 'kosipo'; the timber from each of these is of a slightly different shade and texture and often gives off a pleasant cedarwood smell.

A valuable close-grained rosewood, known on the Continent as 'palisander' and in other places as 'camlai' is derived from various species of *Dalbergia* (Leguminosae), from Indo-China. It is used for inlaid work and decoration, for special wood carving and for veneering. In addition to the dark violet timber from Indo-China, there are dark red or brownish rosewoods from Madagascar. One species in this genus, the *Dalbergia melanoxylon* from Africa, provides an almost black wood and is included in the group of timbers commonly known as ebony. There are many types of ebony, each one from different botanical species but almost all under the genus *Diospyros* or date plum (Diospyraceae or Ebenaceae); they include persimmon trees (*D. virginiana*) with their beautiful edible fruits which grow in Southern Europe but are of Eastern origin.

Whatever the species, ebony is without doubt one of the most precious woods and has been known since ancient times; in Italy in medieval and renaissance times the reputation of this black, close-grained, smooth wood, almost without knots, led to the establishment of a guild of furniture-makers producing furniture from 'les bois des îles'; another guild also existed known as the 'ebonists', or craftsmen doing inlay work and sculpture, often of a high artistic level. It is of interest that despite the flourishing trade in and use of ebony – or rather, of ebonies, for care was taken to distinguish between woods according to the place of origin – nothing was known of the botanical origin of the timber.

Even today, the origin is an important factor when trading in and handling ebonies; there is a black, very hard 'Macassar ebony' from a tree about 130 feet tall (*Diospyros macassar, D. celebica* or *D. utilis*) which is imported from the Celebes in Indonesia to Europe, where it is highly prized; two to three thousand tons of the timber are produced annually.

An ebony from Guinea is obtained from *Diospyros crassiflora* (*D. incarnata*). It comes from the Congo and Nigeria, together with a related species (*D. evila*), a native of Gabon. Angolan ebony, one of the finest black timbers, is obtained from *D. dendo*.

One ebony from Madagascar, derived from *D. perrieri*, has now disappeared because of the irrational way in which the forests have been exploited in the western region of this island. One of the several species of ebony which is now no more than a memory is what is known as the 'bastard ebony' (*D. nodosa, D. angulata*).

From India and Ceylon in the Far East, comes what is known as the East Indies ebony, from whose specific name, *Diospyros ebenum*, is derived the term used indiscriminately for all these black woods which have been made into beautiful statuettes, idols and other objects by native artists, and which has been so widely used in Europe and America for the black keys of pianos.

Tectona grandis and related species of teak woods are native of Indo-China. Because teak is impervious to water, even after prolonged periods of immersion, it has long been used for shipbuilding and for railways, including railway sleepers. It was first introduced as timber for flooring and has more recently been used for furniture, giving rise to the remarkable popularity of the reddish-brown wood in the modern style of furnishing known as 'Swedish'.

In this account it would be impossible to describe or list all the known woods of exotic origin that are used by craftsmen, artists, and the furniture industry: there are at least sixty such woods, with exotic-sounding names such as 'sao' (*Hopea odorata*, Dipterocarpaceae) from Indo-Malaysia; 'moabi' (*Baillonella djave*, Sapotaceae) from the Ivory Coast; 'bang-lang' (a species of

Lagerstroemia, Lythraceae) from Vietnam; and 'iroko' (*Chlorophora excelsa*, Moraceae) from the coasts of Africa.

Tropical plants offer a surprisingly numerous and varied selection of raw materials, from textile fibres and resins to the most obscure substances: natural sources are virtually untapped, despite exploration and the encroachment of civilization.

There is no doubt as to the importance in both temperate and tropical regions of the vegetable materials which are used to supplement wools of animal origin.

Man has always eagerly sought new materials to weave and new dyes to colour these woven cloths, which are used to provide warmth for his body in the colder climates, and for personal adornment in the heat. He also likes to decorate his home, however primitive, with mats and hang-

ings. The development and use of dyes is discussed later, but a few details about textile fibres are necessary, since the development of weaving has always been an indication of the level of civilization attained by a human community. Only when agriculture has developed to a considerable extent does one find that the textile-producing plants are used more widely.

The use of animal fibres preceded that of vegetable fibres, which were rarely known in primitive civilizations. Wool and later silk, were easier to find and spin. At first, vegetable fibres were mainly employed for ropes and cords. These were made by unravelling stems, branches, leaf petioles, or leaf laminae, especially those of monocotyledons, and many palms.

It may have been the demands of fashion that first induced man to seek out fibres that could be

A plantation of Agave sisalana in Tanzania. The famous textile sisal is manufactured from the leaves of this plant

made into fine fabrics and dyed in a wide range of colours. At the same time, coarser but tougher fibres were gradually discovered for various domestic requirements, such as sisal, jute, 'paka', and 'abaca' which were used for carpets, mats, sacks, and string; the plants that supplied these fibres were then systematically cultivated.

One of the outstanding tropical textile-producing plants is cotton (especially the genus *Gossypium* of the Malvaceae family). Although the plant is today cultivated in Egypt, India, Israel, Sicily and other warm places, it is of tropical origin. The most common species are *Gossypium herbaceum, G. hirsutum* from central Asia but originally from Mexico and Guatemala, and *G. barbadense*, originally from tropical South America.

The boll, or fruit, from which the cotton seed is taken has its counterpart in the Asclepiadaceae and Apocynaceae families, from which excellent vegetable silks are obtained. Under this heading come both *Calotropis procera* and *C. gigantea*, from India and Africa, whose product is known as Madar. Some of the best vegetable silks come from species of *Marsdenia*, as well as *Beaumontia grandiflora* from India and *Asclepias syriaca* (*A. cornuti*) and *A. curassavica* from the two Americas. They are also derived from both *Boehmeria utilis* and *B. nivea* (Urticaceae), whose products are respectively the 'green ramie' and 'white ramie', both from China. The lustrous, fine fibres from these nettles are still treated almost solely by hand, since no mechanical system has yet been devised that has proved to be as effective.

Kapok too comes from tropical plants; not from the hair-like coating of the seeds but from the endocarp of the fruits. There are two species of fibre: a kapok from the Far East of Indo-Malaysian origin derived from *Ceiba pentandra*, and the African kapok from East Africa, the plant *Bombax buonopozense* (a species of baobab), both of the Bombacaceae family.

The number of fibre plants which live in the tropical belt is considerable: about fifty mono-cotyledons and even more dicotyledons are used in the textile industry, and together they produce about sixty different types of fibre or exotic fabrics.

The monocotyledons include the bowstring hemp (*Sanseviera cylindrica, S. zeylanica, S. guineensis, S. ehrenbergii, S. longiflora,*) in India, Ceylon, and Africa (Somali Republic and Ethiopia), producing 'Angola hemp'; the aloes – *Agave sisalana, A. elongata, A. salmiana* and *A. zapupe* (Amaryllidiceae) – which produce respectively strong, coarse fibres of sisal (Yucatan and Central Asia), 'hennequen' (tropical Africa), 'ixtle' or 'tampico' (Mexico) and 'zapupé' (Mexico). *Musa textilis* (Musaceae), a banana plant, that furnishes 'abaca' or Manilla hemp (Philippines), and *Phormium tenax* or flax lily (Liliaceae) produces what is known as New Zealand flax for coarse canvas. *Raphia ruffia* and *R. longifolia* palms produce a very strong coarse fibre, raffia, that is used in the making of hats and mats as well as for gardeners' and fruit growers' string.

Vegetable 'horsehair' has been used for centuries as filling for mattresses, chairs and so on; this is derived from palms of the genus *Chamaerops* (*C. humilis, C. hystrix, C. fortunei, C. ritchiana*).

The dicotyledons include species that furnish other fibres for coarse fabrics, such as jute or nalta from *Corchorus olitorius* (Tigliaceae) of India; 'sida', or Queensland hemp from *Sida rhombifolia*, a member of the Malvaceae family from tropical Asia, Central Africa and Oceania; 'guama' from *Hibiscus quinquelobus*, another member of the Malvaceae, the 'hafotra' from various species of the genus *Dombeya* (Sterculiaceae) from Madagascar; and 'sunn' or Calcutta hemp from *Crotalaria juncea* (Leguminosae), from northern Asia.

Tropical species are also used to produce cellulose for paper, for example the papyrus from Gabon and the various bamboos from the former Indo-Chinese territories, as well as the agave, the 'palma dum' and certain species of the genus *Stipa*, including esparto grass.

Poisons and magical drugs

The use of magic for therapeutic purposes is universal, and it still forms part of the heritage of many tribes dwelling in the African, Asian, and American tropics. The administration of ointments, cataplasms, or potions is often accompanied, in primitive societies, by convulsive dancing, ritual gestures, and fires. These ceremonies are supposed to enhance the potency of the 'medicine', and the victim or patient dies or recovers accordingly. There is no doubt that many of the innumerable products of primitive herbal medicine and experimental drugs were and are still valid. Balsams and plant juices with curative properties constitute the basis for the majority of pharmaceuticals that are used in Western society, with the exception of synthesized drugs.

The magic of the healers, however, is generally only as powerful as the drugs they administer. Undoubtedly, there is a greater quantity of poisons and medicaments among the tropical plants than in those of temperate or cold zones. The whole atmosphere of these torrid climates seems to be conducive to magic and mystery, but the deadliness of some of the plants and animals is a harsh reality. In the jungle it is by no means a thing of the past to find men firing poisoned arrows, nor is it so rare an adventure to encounter poisonous snakes (whose bites are often fatal), among the dense network of lianas and ferns. The spiders of the hot deserts, despite their terrifying appearance, are more venomous than deadly.

The tropics, therefore, are the kingdom of poisons: there is, for example, a liana known to the natives as 'urutungo guaiajana' (literally, flower of the jaguar), which bears pink-violet prickly plants like horse chestnuts, and whose thorns inject a deadly poison into the paws of any incautious wild beast, causing it to die an agonizing death.

According to those who specialize in toxicology, some of the most poisonous plants are to be found among the *Strophanthus*, including *Strophanthus gratus* and *S. kombe* (Apocynaceae) from East Africa and Madagascar, from which the strophanthosides are still derived. These substances are derived from the seeds, and have an action on the heart similar to that produced by digitalis. Their properties have been known since ancient times; the Somalis call them 'cubaio'. The first seeds were imported into Europe in 1875 by the explorer Griffon du Bellay.

The toxic strophanthins (some species are quite harmless) were and still are used as poisons for arrowheads. To prepare the poisonous liquid, the Indians remove the husks and crush the seeds to a pulp between two stones; they then add water, saliva and poisonous vegetable juices, with fragments of flesh and putrefied animal organs. It is known that birds wounded by arrows dipped in this poison fall as if they were struck by lightning. Larger animals die after running only a few yards, and elephants will die after travelling a few miles if the arrows penetrate deeply.

Native Africans used to employ this murderous poison against enemy tribes; they would scatter splinters of wood that had been sharpened and dipped into the poisonous liquid on the tracks through the trees and bushes, knowing that sooner or later they would penetrate the bare feet of unfortunate travellers and cause their death.

There are various types of *Strophanthus*: some are bushy or twiggy shrubs and some are lianas; they are to be found in abundance along the sunny areas at the edge of forests.

Equally poisonous are the substances derived from the tanghinias, notably *Tanghinia venenifera*,

The kapok tree Ceiba pentandra, one of the Bombacaceae from tropical Africa

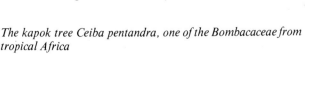

100

Left: An aloe in flower.
This plant species has
been known for its
medicinal properties
since ancient times

a native of East Africa; their poisonous juices, whose effects are similar to that of curare and strophanthin, are used in small doses by the natives during their mystic ceremonies, especially in Madagascar. The poisons from these two species have been employed since time immemorial to render the arrows capable of fatally wounding their victim.

The same can be said of the juice derived from the bark of *Erythrophloeum cuminga* (Leguminosae) known to the natives of East Africa and Madagascar as 'telei' or 'tali', like the *Erythrophloeum guineense*.

The curares are mainly strychnine-based complex drugs that are taken from secret native formulae transmitted from ancient times to the present day, particularly among the tribes of the Amazon basin and the Orinoco in the northern part of South America. These plants are highly poisonous. A feature is that they may often be innocuous when taken orally but are poisonous when they enter the bloodstream, to such an extent, indeed, that minimal doses entering the body through a tiny scratch may cause instantaneous death.

This extreme toxicity is accounted for by the action of curare on the nervous system. Messages travelling in the body pass from one nerve to the next across a small gap, by means of a chemical substance, and what curare does is to compete with this chemical, blocking the transmission of nerve impulses. Although deadly, this property of curare has been extremely useful to scientists studying the working of the nervous system of animals, especially when using single nerve fibres in isolation in the laboratory. It is also a curious feature that convulsions do not accompany curare poisoning, as is the case with true strychnine, which has a tetanizing effect on the muscles.

The natives of Amazonia use strychnines taken from various species of *Strychnos* (the Loganiaceae family) but know how to add many other ingredients and use special methods of preparation which profoundly alter the strychnine effect of the poisons.

Returning to the origin of the word, the term 'curare' appears to be a corruption of certain names given by the Indian tribes of Amazonia to the poisons they compounded, especially the term 'uraréry' which means 'liquid that kills birds' and 'caruchi' or 'mavacuré', words from the local Caribbean language having the same meaning.

The discovery of curare by the Europeans was the result of the Spanish conquest of the territories of the New World; the Spaniards, indeed, were the first unfortunate European victims of

Sugar cane (Saccharum
officinarum) near Lae
in New Guinea

the poison. Today curare is used by the natives only to poison hunting arrows and is prepared from a considerable number of vegetable species, using different methods and in forms that differ in their lethal power.

A comprehensive list of 'magic' soporific and hallucinatory drugs would be exceedingly long: it would include certain fungi of the genus *Psilocybe* that are used in Mexican Indian religious rites; opium derived from the poppy (*Papaver somniferum*) from the East, from which several drugs in the morphine group are obtained; and Indian hemp (*Cannabis indica*) known as 'hashish', a 'soft drug', which is illegal but is used in many countries of the world.

There is a drug which plays a dual role, the coca, which is the source of a well-known

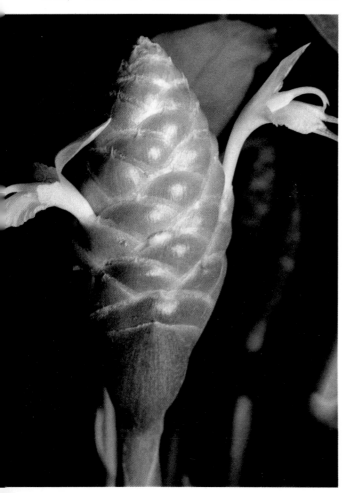

alkaloid, cocaine. Although cocaine lends itself to abuse, it is highly valuable in the field of medicine. It is derived from the leaves of *Erythroxylon coca* (Erythroxylaceae), a native shrub of Peru. The Amerindians in this region were the first to discover the properties of the leaves whose use was once the sole prerogative of the Inca chiefs.

Coca leaves are chewed slowly to form a bole which is retained in the mouth, or can be crushed into various kinds of pulp for mastication. A little slaked lime or plant ash from certain Cactaceae or *Chenopodium quinoa*, a member of the Chenopodiaceae, and other floury substances are added to form what is known as the 'sit-tikkchira-carapari'.

The local tribesmen chew these substances with a high proportion of coca leaves to help them withstand physical fatigue when undertaking long treks, as well as to overcome the feeling of hunger. Unfortunately, the need for coca gradually becomes an escapist craving and finally turns into a fatal drug addiction.

In medicine, however, cocaine has been very useful. Moreover it illustrates how plants can provide not only a single drug, but also the basis of a family of drugs; many drugs are now manufactured that are similar to cocaine but which lack its disadvantages.

Above: A sprig from t camphor tree (Cinnamomum camphora), a laurel from Japan and Formosa. Left: Zingiber officinale, ar Indian ginger plant, bursts into flower

Drugs, dyes, and spices

In theory, there is no strict dividing line between plants that are poisonous and those that are useful in providing us with drugs, dyes, foodstuffs, and so on, because the difference lies in the strength of the active chemical compounds in the plant tissues and the amounts which man normally encounters. The plants described in the previous chapter are poisonous, it is true, but many have been directly or indirectly valuable in medicine.

We will now go on to consider species that are less toxic – but which, of course, are not harmless if used carelessly or in the wrong quantities – and from which man obtains a wide variety of valuable products. Energy food plants (vegetables and fruits) will not be included here, but we will discuss those species that provide spices, flavours and drinks, such as coffee and cocoa, others that are sources of drugs and pigments, and some that have been used for several different purposes.

The first to consider is quinine, a very important drug extracted from the bark of numerous trees belonging to the genus *Cinchona* (Rubiaceae). The use of this important drug and its derivatives is very widespread, especially against malaria, although it is not known how or when it reached Europe for the first time. The natives of Peru had been familiar with the 'quinquina' bark for hundreds of years, but they never realized its true significance. In the dialects of certain Indian tribes, 'kina' means 'bark', and the expression 'kina-kina' is a superlative, meaning 'bark of barks', that is, the best and most valuable of them all. From this come the Spanish language corruptions 'china-china' and 'quinquina'.

History relates that in the first half of the eighteenth century, a Jesuit missionary and an official from a village in Peru were saved from a terrible fever by a healer from the Indian village of Malacota who used a magic powder derived from the crushed bark of cinchona. Eight years after this cure, which was then considered to be miraculous, a similar experiment was conducted on the person of the Spanish vice-reine, the Countess Chinchon. Quinine powder, which had first been known as the 'powder of the Jesuits' was thereafter called the 'powder of the Countess'. Under this name it was introduced into Spain, primarily to Alcala de Hénarés, the birthplace of the Jesuit order.

Once it had become a valuable medicine, quinine spread to Italy, being originally introduced into religious communities, as in Spain. The original generic name was *Chinchona*, in memory of the Countess.

Although the powder was known to have cured fevers in most parts of America and Europe, very little was known of the original plants. These were finally discovered by La Condamine who was sent out to survey an area in southern Peru, or rather by one of the members of his group, J. de Jussieu, a great botanist whose brothers were the famous Parisian naturalists, Antoine and Bernard de Jussieu.

The discovery was published in 1739, and after that many explorers sought, studied and described cinchona. There are about twenty allied species which grow wild from Venezuela to Bolivia, including Colombia, Ecuador, Peru, and probably the northernmost part of Chile.

The trees have many branches whose elliptical leaves are placed opposite each other. There is an inflorescence at the tip, which takes the form of a cluster or corymb made up of numerous five-pointed corolla-shaped flowers, which range from white to bright pink in colour.

There is no doubt that quinine is one of the most useful medicinal plants that the tropics have provided for man's use, but there are many others which combine fragrance and therapeutic value. Among these is camphor, and the spices include cinnamon, cloves, nutmeg, and pepper.

Camphor is usually derived from *Cinnamomum*

*The green berries of
Coffea arabica turn red
as they ripen, and the
mature seeds that they
contain are roasted to
produce coffee*

camphora (also called *Laurus camphora*), Lauraceae, but a similar product, known as camphor of Borneo or Malaysia, may be produced from *Dryobalanops camphora* (*D. aromatica*, Dipterocapaceae).

With the first type, the true camphor, the wood of old camphor trees is distilled. These trees used to abound in the Far East, southern China, Japan, and especially in the island of Taiwan. They are strikingly decorative, reminiscent of the lime in appearance, but evergreen, with oval, pointed leaves. In their areas of origin they grow at altitudes of up to 2,000 metres (6,500 feet), but in Europe they flourish only around the Italian lakes. They often live to an age of a thousand years and their trunks grow to a diameter of no less than two metres (six feet). This fragrant drug has been used in heart diseases by the peoples of the Far East for many centuries. However,

camphor was unknown by the Greeks and the Romans, and it appears that it was introduced to the West by the Arab prince, Imru-Al-Kaisu in the sixth century.

In the early part of the ninth century camphor was obtained by making deep incisions into the bark of camphor trees; a little later, the refining process was developed. The Arabs traded in camphor as a rare perfume used only by the wealthy and it was Aetius, a medieval doctor, who first used the substance in medicine.

The other species of camphor, derived from *Dryobalanops camphora* (*D. aromatica*, Dipterocarpaceae), and known as a camphor of Borneo or Malaysia, is produced spontaneously from the wounds or holes in adult trees bored by the larvae of a coleopterous parasite.

Cinnamon, too, is derived from a member of the Lauraceae, and here a distinction is made

between several varieties of the drug: the principal form comes from Ceylon or Malabar and is derived from *Cinnamomum zeylanicum* (*Laurus cinnamomum*); another kind is the cinnamon from China and Burma (*C. cassia* and *C. obtusifolium*). Cinnamon is also found in Annam, where it is known as the 'royal' variety, and in Malaysia, South Vietnam, the Seychelles, and other places. Cinnamon is the bark of the *Cinnamomum* tree; when it is dried it rolls up into small cigarette shapes that are sandy-brown in colour and have an agreeable, aromatic smell.

The cinnamon trade began in Ceylon several centuries ago, probably in 1518, when the Dutch seized the vast tropical island from the Portuguese. From 1770 onwards, cinnamon trees were cultivated there, the first planter being a settler named De Koke, who founded the famous 'cinnamon gardens', or 'Jardins de Cannelliers', which are still open to visitors to Colombo.

The medical use of cinnamon as a circulatory and respiratory stimulant and as a vasoconstricting agent is well known, and it is still used as a spice in cooking and as a flavour for beverages.

The clove is equally and justly well known. The flower buds from the clove tree, *Eugenia caryophyllata*, a member of the Myrtaceae family, are picked before they are open. The pungent scent, reminiscent of the geranium to which the tree is related, has earned it the names *Caryophyllus aromaticus* and *Syzygium aromaticum*.

The tree is a native of Madagascar and the Moluccas (Spice Islands); it has beautiful foliage and grows to a height of 10 to 15 metres (up to 50 feet). Its branches have tough, leathery leaves and it produces a terminal inflorescence in the shape of corymbs, each made up of about twenty small, pink flowers.

Europeans first encountered this tree in the island of Ambon in Indonesia, although the essence had been extracted for centuries in the Far East. However, since the clove has been used since time immemorial, it is difficult to establish its geographical origin; generally it is to be found in the hot and humid climates of the immense archipelago in the tropical oceans of Asia.

The Chinese used the clove long before the Christian era, and from the earliest times the Hindustanis considered the substance to be an essential ingredient in the flavouring of food. Their formulas for spices are still famous today, and have been handed down for thousands of years.

The fact that the clove was used in antiquity has been confirmed by its discovery in many sarcophagi. The Phoenicians introduced the spice to the Greeks and Romans, and it was used in France from the time of the Crusades. The Portuguese navigators discovered what were named the 'Spice Islands' during the fifteenth century, and they were also responsible for importing the clove into Europe on a regular basis.

Another substance of tropical origin that has long been in common use in Europe is the nutmeg, the seed of *Myristica fragrans* and that of two related species, *M. moschata* and *M. officinalis*. The trees are evergreen and when fully grown reach heights from 8 to 10 metres (up to about 30 feet). They produce pendulous, pear-shaped drupes which split in two when mature. They contain a large shiny seed with a bright red outer covering (aril).

These aromatic seeds come from the Spice Islands (Indonesia), and they were not known to the peoples of antiquity; it was only in about the eleventh century that the Arabs used them as

109

aromas, and Constantine the African cites them as a rare and expensive substance. When the Dutch took the Banda Islands from the Portuguese they organised a spice monopoly (cinnamon, cloves, etc) that included the nutmeg; during this period, most of the trees of *Myristica fragrans* were destroyed. In 1769 the botanist Poivre came into possession of a few young plants and took them to the Mascarene Islands, whence they spread to Mauritius,

Cocoa pods (Theobroma cacao), one of the Sterculiaceae now cultivated in many tropical regions

becoming a profitable crop. Today nutmeg is used as a spice in its own right and also in the preparation of a special butter, as well as for mace, which is produced from the dried outer casing of the drupe stone.

One of the most widely used and important spices is pepper (*Piper nigrum*, Piperaceae). The plant is a sub-ligneous liana, a native of Malaya. It is grown all over Indo-China, especially in Cambodia and South Vietnam. The peppercorns are the berries, which are gathered and dried before being separated from their fleshy outer layer to produce what we know as white pepper.

Together with the nutmeg, cinnamon and the clove, pepper was one of the main spices which acted as a magnet to the navigators in the Middle Ages, inciting them to embark on long voyages of exploration to the East. The history of pepper is, therefore, inextricably associated with the history of all those spices and drugs that were such prized commodities for so many centuries.

There are many pepper substitutes, such as the fruits and other parts of the cubeb (*Piper cubeba*), the matico, or *Piper angustifolium* (*Artanthe elongata*), the betel (*Peper betel*), the 'kawa-kawa' (*P. methysticum*), the long pepper (*Chavica officinarum*), and the jaborandi (*P. jaborandi*).

In all these species it is the berry which is used, with the exception of the matico, whose leaves are employed. The matico comes from South America (especially Bolivia), and it is used as a hemostatic agent in medicine.

Other plant substances are even more important than the spices, not because they have nutritional value or are used for flavouring food, but because they are the source of such stimulating drinks as coffee, tea, and cocoa.

Coffee (*Coffea arabica* and other species from the same genus) is a shrub of the Rubiaceae family, whose beans are roasted to produce the beverage. The plant, a native of Africa, has paired leaves or leaves in whorls of three; they are tough and leathery with wavy edges, and are a rich green colour, while the white, sweet-smelling flowers have four or five petals and are grouped in dense clusters growing from the axil. The fruit is a small drupe with a fleshy mesocarp containing two horny albumen seeds. There are about twenty species.

Although the native land of the best coffees is Africa, certain less prized species and others with no economic value are to be found in western Asia and especially in Arabia.

They do not grow wild in the continent of America, but the best species and varieties have been cultivated in Brazil and Mexico for a long time. Only one species of coffee was known

before 1790 (*Coffea arabica*); it grew wild in the north-eastern part of Africa. Other species were later found in Kenya, Malawi, Gabon, Mozambique and other countries.

Coffee was first imported into Europe from Arabia through Venice in 1640. The botanist Charles d'Ecluse, known as Clusius, was the first westerner to study coffee beans, and Prospère Alpin reported that he had admired the plant in a Cairo garden in 1592. By 1644, Jean de la Roque had taken coffee from Constantinople to Marseilles where it became popular, and Louis XIV

tasted it for the first time in Paris in that year. It was in Paris that the first 'cafés' were opened in 1671 and 1672, and the public was able to taste this new beverage.

Scholars and critics were not all happy about the introduction of this new substance or the drink derived from it. In 1679, Colomb, a young doctor from the Aix faculty of medicine, on the advice of his professors, wrote a thesis entitled *Savoir si l'usage du Café est nuisible aux habitants de Marseille* ('To find out whether the use of Coffee is harmful to the inhabitants of

A North African date palm, Phoenix dactylifera, in full flower

112

Marseilles'). Finally, it is of interest that Louis XIV, Fontenelle, Voltaire, and Napoleon I were coffee-lovers, while Frederick the Great and the English resisted it, the latter remaining faithful to tea.

Tea is made from the cut and dried leaves of a tropical shrub, *Thea sinensis* (*Camellia sinensis*, Ternstroemiaceae), a native of eastern Asia. In the trade a distinction is made between 'black' tea, produced from leaves which have been processed and fermented to alter the original chemical composition, and 'green' tea, which is roasted while still fresh, and scented with aromatic plants.

The history of tea is lost in antiquity: it has been discovered that the Chinese were using it more than 2,500 years before the Christian epoch, although it is not clear what the real reasons were for their use of the beverage. According to legend, it was thought to be of divine origin.

The Chinese and the Hindus both claim the honour of having first brewed the 'cups that cheer but not inebriate'; the Japanese have known it for more than 1,200 years. Marco Polo tasted tea during his journey to China and praised its taste. In Europe, tea has been known since the beginning of the seventeenth century when it was imported by the Jesuit missionaries into France. However, it was popularized by the Dutch who not only established the tea trade but organized the bartering of tea and another valuable vegetable substance, sage. It was the English, however, who were responsible for the world-wide popularity of tea, attributing tonic and invigorating qualities to it.

Cocoa should also be mentioned briefly, for it is undoubtedly one of the most important tropical products. It is the main ingredient of chocolate, and is derived from the fruits of the cacao tree, *Theobroma cacao*, and other related species (Sterculiaceae). These small trees grow in Mexico and other parts of Central America, and are now cultivated in many equatorial regions. The tree has a straight trunk, with slender, greenish branches, and large, elliptical leaves.

The flowers are borne singly, or are grouped in small clusters of two to five, rarely more. They grow from the axil of the leaf after it has fallen, either on the trunk itself or on the larger branches – a phenomenon known as 'cauliflora'. The flowers are yellowish-white with pink or purple spots.

The fruit has a short peduncle; it is oblong in shape and grows to a length of 10 to 20 centimetres (up to eight inches). It is rounded at the base where the large peduncle is attached, and thinner at the apex. It hangs downward when ripe. The fruit is ribbed in section and roughly pentagonal, growing to a diameter of six to ten centimetres (up to four inches). It has several internal septa or partitions with flesh in between. The outer skin is red or yellow; it is hard, thick, and sometimes lumpy. The pericarp is fleshy on the outside and fibrous on the inside, and contains 20 to 40 large seeds coated with a sticky sugary pulp. Cocoa beans may measure one to two centimetres (less than one inch) and contain two large cotyledons, or seed-leaves.

Legends abound on the subject of cocoa; one of them defines the 'cacahuatl' as being of divine origin. This was the belief of the Aztecs who used it before the birth of Christ; perhaps it was this legend that inspired the early botanists to name it 'theobroma' or 'food of the gods'.

During their conquest of the American tropical lands, the Spaniards were introduced to cocoa and chocolate. In about 1625 they brought back specimens to Europe. In France, the prized beverage was served for the first time at the marriage of Louis XIV and Princess Maria Teresa, the King of Spain's daughter. In 1659, the king granted a production monopoly by conferring his personal favour on the new product.

Chocolate is produced from cocoa beans from which the outside pulp has been removed. The beans are fermented, roasted and crushed to a thick paste that is mixed with sugar and flavourings such as vanilla.

The few drugs and spices that have been briefly described here by no means complete the list. Cola, ginger, sandalwood, cardamom, vanilla, and many others have not been mentioned. There are, however, two other groups of valuable substances to consider: balsams and dyes.

Myrrh and bdellium belong to the first group, being derived from *Commiphora abyssinica* and *C. molmol* (Burseraceae) respectively. These plants are native to India. The various 'elemi' gum resins come from the genera *Commiphora, Canarium, Protium,* and *Elaphrium* (Burseraceae), and widely differing tropical regions: from Indo-China to the Antilles, and from Brazil to Angola and Madagascar. Tolu balsam comes from *Toluifera balsamum*, a member of the Leguminosae from Venezuela and Colombia, and finally there is copaiba, a balsam obtained from certain species of *Copaifera* from Brazil and Venezuela.

The following resins should also be mentioned: dammar, derived from the Dipterocarpaceae of Borneo and the Malay archipelago, mainly from tree plants such as *Hopea adorata, H. micrantha, H. thorellii,* and *Anisoptera costata*; sandarach, the powdered resin of a Moroccan conifer, *Thuya articulata*; and benzoin, popularly corrupted to benjamin and gum benjamin, which is used for fumigation as it was thought to provide a method of pest control. It is derived from *Styrax benzoin* and the related *S. tonkinensis* (Styracaceae) which grow in Laos, Thailand, and Sumatra.

The ingredients used in the making of paints include the copals of Malaysia and Makasar (*Agathis australis*, of the conifer family, New Zealand), those of Zanzibar and the Congo (*Copaifera demensii*, Leguminosae), and the various forms of kino from Malabar, Gambia, the Antilles and Bengal, obtained from Leguminosae (*Pterocarpus sp.*), Myrtaceae, Polygonaceae, Rhizophoraceae, and so on.

Despite the usefulness of the tanning materials derived from wood and bark, and the substances used for dressing leather and staining woods, there is only space to mention a few colouring matters and precious dyes of the East. One of these is indigo, whose beautiful deep violet-blue dye is used in many types of artists' paints, calico and fabric dying, and in chemicals. It is obtained from the indigo plants, *Indigofera tinctoria, I. anil* and other related species, of which all are Leguminosae commonly found in America (especially Central and South America), Indo-Malaysia, Africa (Senegal and the Sudan) and Arabia.

Indigo is extracted from the leaves of the plant and is processed by reduction and deoxidation of a glucoside, called indican. It originated in India, although it was known to the ancient Egyptians: mummies dating back from before 5,000 BC were wound with fabrics that had been dyed with indigo. There are many known indigo plants today and they constitute an important crop in tropical regions.

Another colouring substance, haematoxylon, is a dye that may be reddish, violet, or blue; it is obtained from logwood, the wood of *Haematoxylon campechianum*, another member of the Leguminosae. The tree comes from equatorial America, and grows to about 32 to 48 feet (up to 15 metres); it has compound leaves and spiny stipules. In its natural state the wood is yellow, and the violet-black pigment is obtained by treating the timber with ether and ammonia.

Other woods used as dyes include Brazil wood (*Caesalpinia echinata*) from Brazil, producing a brownish-red pigment, sapan wood (*C. sappan*) from southern Asia, the red wood of Jamaica, and red sandal wood (*Pterocarpus santalinus*) from Malaysia and Ceylon which produces a red dye. Gamboge is an intense yellow colouring widely used in painting; it is obtained by making deep incisions in the trunk of a small tree (*Garcinia hanburyi, G. gutta*) which belongs to the Glusiaceae (Guttiferae) family, and which grows in the humid forests of Ceylon, India, Cambodia, and Thailand. In its natural state it is a yellowish resinous sap that becomes viscous on contact with the air. It is poured into bamboo canes and later decanted by the application of gentle heat and made to set in small stick shapes.

The coconut palm, Cocos nucifera, *in Jamaica. Under cultivation, a single tree may yield between 100 and 200 coconuts a year*

Fruits of the world

Man has always been a lover of fruit. Right from the earliest times of his history, when hunting and gathering were the only means of obtaining food, through to the present day when crop production, marketing and buying are highly advanced, the many and varied fruits have been among the mainstays of his diet.

The reason for this, according to some biologists, is that man needs vitamin C, and fruits can provide it. Very few other animals need this vitamin, and it is notable that all of them like fruit: examples are monkeys and the 'fruit-eating' bat. Whether or not this does explain man's fondness for fruit, and indirectly his sweet tooth also, there is no doubt that fruit is a highly prized type of food.

The great majority of fruits consumed throughout the world are of tropical origin. Temperate

Below: A banana plantation. Above right. The strange flowers of the banana tree. Below right: Durian fruit (Durio zibethinus) in a Thai market

lands saw the origins of many types of apple and pear tree, the cherry, hazel nut and chestnut trees, and many vines and fig trees. However, in addition to the obvious examples of the coconut, date, banana, pineapple and prickly pear, many other fruits are of exotic origin, including the citrus fruits (tropical Asia, China, Malaysia), the peach (China and Iran) the Chinese date-plum or Japanese persimmon, the plum (Asia), the apricot (India, China, and possibly Egypt), the Japanese medlar, and the pomegranate (Transcaucasia, Iran, or Africa). These trees have now been adopted in more temperate zones, and this sometimes happened in prehistoric times. Other fruit-bearing trees that seem rare to the average American or European do in fact form the staple diet of the people of the countries in which they are indigenous.

The first of these is the cashew-nut tree or acajou, the *Anacardium occidentale* (Anacardiaceae). This is a smallish tree of tortuous appearance, with oval leaves and small flowers on terminal spikes. The fruit is a large achene similar to an overgrown kidney bean in shape, and is coated with a layer of flesh topped by a peduncular, pear-shaped swelling; it is yellow and red in colour. Inside the achene is a stone that is also edible, known as the acajou, which is eaten either raw or toasted. The cashew nut is grown in many regions of India and tropical America and large quantities are exported all over the world for use in cakes and sweets.

Another delicious tropical fruit is the avocado pear, sometimes known as 'alligator pear', from the plant *Persea gratissima* (*Laurus persea*, Lauraceae), a tree from Central and South America. The fruit is about the same size as an ordinary pear and ranges in colour from yellow to green, green with a tinge of red or even dark violet. Its thin, delicate, mottled skin protects the greenish white flesh, which is delicately buttery in flavour but does not have a strong taste of its own. It has a large oval stone that is brown in colour. The avocado has a high content of fats, proteins, and vitamins, and is an excellent food; it is widely used in the diet of the

inhabitants of many tropical regions. The flesh is also used to make a vegetable butter. The fruits of the *Persea drimyfolia*, known locally as 'aguacate', are also excellent and are used in the same way.

A very important tropical fruit is the paw-paw, from the *Carica papaya*, a member of the Passifloraceae. It is not certain whether this plant originated in Mexico or in Pacific islands such as the Spice Islands. *Carica papaya* is a beautiful, fast-growing tree, with a tall, straight, bare trunk and branches only at the top, looking like a giant upright mop. The large, palmately parted leaves have a long petiole. The male flowers are borne on branching racemes growing from the axil while the female flowers are in corymbs of two or three flowers and are white or greenish. The fruit is a large pyriform-ovoid berry rather like a melon in shape; it is orange-yellow in colour and contains many large seeds. The fruit is generally eaten raw like melons and bananas, since it is sweet and fragrant, although it may be baked or fried. The paw-paw's main ingredient is a pepsin substance, so that its flesh is almost a predigested food, and this is the reason why it is eaten at the end of meals.

Another delicious but very perishable fruit is the mango, obtained from the *Mangifera indica*, another of the tropical Anacardiaceae. The tree, which originated in Indo-Malaysia, is a beautiful

Mangoes (Mangifera indica), an important source of nourishment for many millions of the world's inhabitants, pictured in a market with another vital food source, coconuts

The fruit of Pandanus ectoria in a New Guinea plantation

The fruit of the bread-tree or monkey-bread tree (*Artocarpus incisa, A. communis, A. incisifolia*, all of the Moraceae) is native to Polynesia. It is syncarpous, spherical or globose-clavate, and grows to the size of a melon or even larger, weighing two to six and a half pounds. The skin is warty and reticulate, while the white, close-grained flesh resembles bread. The flesh may be eaten raw but it is undoubtedly better when fermented and cooked. The fruit has 70 to 100 large pips, as big as chestnuts, which are also good to eat, being sweet and floury when roasted.

Artocarpus integrifolia is another related plant growing in India. Its fruits are larger than the bread-fruit, and may attain a weight of up to 26 pounds.

Certain varieties of bread-tree produce seedless fruits, while in the case of others the fruit is picked before it is completely ripe, and is then baked or toasted before eating, thus resembling baked potatoes in texture and taste.

The Tahitians prepare a highly nourishing dish by fermenting the flesh of artocarpus fruit which they call 'popoi'; it tastes of cheese and can even be stored for a period of years.

There are innumerable tropical fruits, but of greatest interest are the San Domingo apricot, similar to the mango but obtained from *Mammea americana* (Guttiferae); the American 'sugar-apple', a fruit of *Mammea aquamosa*; pecan or hickory nuts; the delicately flavoured lychees or litchi from China; brazil nuts from *Bertholettia excelsa* (Myrtaceae) whose hard husks can be cracked only with the strongest of nutcrackers; the fruit of the *Aberia caffra* (Bixaceae) which are like pear-shaped plums; and the fruit of the *Feijoa sellowiana* (Myrtaceae) from Uruguay and Brazil, whose flowers are also edible.

Other minor fruits include the custard apples from the *Asimina triloba* (*Anona triloba*, Anonaceae), a native of North America, which are very similar to bananas; the tamarind (*Tamarindus indica*, Leguminosae) with its rather sour flesh; the durian from Malaysia and the 'lecythis' from a Brazilian myrtacea, a woody fruit the size of a melon with many edible seeds.

one, and it bears succulent, fleshy fruit that have a slightly resinous taste, ranging in size from the equivalent of an egg to that of a man's head and sometimes growing to weigh over six pounds.

Another type of mango, from the tree *Mangifera elephantina*, is also excellent. It should be eaten newly picked, and is so delicate that it does not travel well. In mango-growing regions, the fruit is fermented to make an alcoholic drink.

The large fruits of the baobab (*Adansonia digitata*) are also edible, as are those of the bread-tree. The former are as big as pumpkins and are elliptical-oblong in shape, and yellow ochre or greenish in colour. They contain a pleasant sweet-sour flesh.

119

Tropical cash crops

There is no doubt that the cultivation of fruit-bearing plants and the demand for certain beverages (coffee, etc) have resulted in a standard of agriculture in certain tropical regions that very nearly approaches the level achieved in most European countries and in America, although the development has been very recent.

Nonetheless, the sudden agricultural and industrial transformation of tropical land has in many cases upset the natural order that has reigned there for thousands of years, often irreversibly changing the climate and ecology. Soil that has been exploited for a quick return over a brief period rapidly becomes farmed out, and no further crop can be grown.

It is essential, therefore, that the natural wealth of these countries be protected against economic destruction. There is a need for research on agricultural and industrial methods and administration that will provide true and lasting benefits for the indigenous populations, protecting them from the risk of land impoverishment. The warning was sounded decades ago with the phrase 'Africa, the dying land'. The felling of great expanses of forest for agriculture has denuded extensive tracts of land in which the rain, no longer tempered by protecting boughs, has eroded the soil and disastrously washed away the nutrients contained in its upper strata.

The disappearance of the forest – and therefore of the undergrowth and the microflora – from the biological community, has seriously affected humus production. At the same time, the hot sun raises the temperature and decreases humidity with the adverse result of deoxidizing and mechanically desegregating the soil itself. In short, all these developments have caused irreversible harm to the soil which is neglected altogether once its agricultural properties have been exhausted.

It is true that the vast expanses of land that could be put to agricultural use provide ample space for man in his inexorable search for new soil to farm, but by acting in this way he leaves behind him an increasing area of potential desert. Although the burden of greed will weigh heavily upon humanity for centuries to come, tropical countries have shown that great wealth can be derived from proper utilization of the soil.

Since Africa, America, Asia and Australia were first colonized, the incoming people sought out animal and vegetable resources and selected from the fauna and flora the species that were most suitable for breeding or cultivation on a profitable basis. Only later were species introduced from other regions, often from other continents, sometimes with excellent results.

A very important member of the Gramineae is rice, which originally came from India and has been cultivated in China for 5,000 years. It was brought from China to Japan in the seventeenth century BC, and finally returned to India as a cultivated cereal, to Ceylon and Egypt in the seventh century BC, and to many other tropical countries as well as the temperate zones of Europe. According to fairly recent statistics, the production figures for rice are as follows: Asia, 126,900,000 tons; America, 2,640,000 tons; Africa, 1,520,000 tons; Europe, 1,180,000 tons; Oceania, 36,000 tons.

Coffea arabica and other species of the same genus, are native to certain regions of Africa and Arabia, whence coffee was introduced to America (especially Brazil) and certain areas of Asia. Although it is not possible to give current production figures for these regions, the exports over the decade 1930–1940 give some idea of the quantities involved. Briefly, these are: South America (Brazil, Colombia, Venezuela, Bolivia), 1,146,200 tons; Central America (Salvador, Guatemala, Mexico, Costa Rica, Nicaragua,

A plantation of coconut palms (Cocos nucifera) at Moorea (Polynesia)

120

Haiti, Cuba), 240,000 tons; Asia (India, Indonesia, etc), 103,000 tons; Africa (Ethiopia, Tanzania, Madagascar), 40,000 tons.

More than 6,750,000 acres of land are devoted to the cultivation of coffee, 3,750,000 acres in the Brazilian state of São Paulo alone.

Tea (*Thea sinensis*) undoubtedly originated in China and then spread to India, Ceylon, Java, Japan, Kenya, Malawi and the USSR (Georgia), all of them regions where the agricultural methods are far more advanced than in the country of origin. Listing the producer countries in order according to the area of land on which tea is grown, India has 82,500,000 acres, Ceylon 565,000 acres, Indonesia 480,000 acres, Taiwan 110,000 acres, Japan 73,000 acres, and territories in what used to be Indo-China – Laos, Cambodia, Vietnam, etc – 25,000 acres. No reliable figures are available on China.

Cocoa (*Theobroma cacao*), a plant that originated in Central America and Mexico, was introduced in more recent times to hot regions; it is fairly difficult to grow, even though it has now spread to most tropical regions as new varieties that are better adapted to each specific area have been evolved by selection.

There are three main groups of these varieties. The 'Creole' or indigenous cocoa is a tree which is very vigorous and produces the best commercial fruit. It is extensively grown in Nicaragua, Trinidad, St. Thomas and Caracas, with two principal varieties, 'amarillo' (yellow) and 'colorado' (red).

The second group consists of the 'foreign' type of cocoa, in other words plants that are not natives of the place where they are cultivated. The tree is hardy and the fruit contains large, cylindrical seeds and is of medium quality. The fruit is classified into three types: 'warty', which is orange-yellow in colour, 'simple', also orange-yellow, and 'melon-shaped', yellow or red. The word 'calabacillo' is used to define the third group, which produces small, flat seeds whose fruit is of lower quality and more bitter in taste.

The leading producers of cocoa today, in decreasing order of importance, are as follows:

Ghana, Brazil, Nigeria, Ivory Coast, Cameroon, Togo, the Dominican Republic, Ecuador, Venezuela, South Domingo, Colombia, Costa Rica, Trinidad, Fernando Po, Mexico, St. Thomas and Panama.

The Gold Coast produces about 300,000 tons a year, and its cocoa is of far higher quality than that of any other country. It is curious that while coffee is a native of Africa but is produced mainly in America, cocoa is American in origin and is chiefly grown in Africa.

Almost all the tropical fruits described on the preceding pages are now grown intensively, from the cashew nut to the paw-paw, from the mango to the San Domingo apricot, without counting the coconut and date palms and the banana tree whose production has become such a vital factor in the economy of many tropical nations.

The fruits of the banana tree have become so common that they are to be seen in the greengrocers' shops and the restaurants of every town and village of the world, from the Arctic circle to the crowded cities of America and Europe. Bananas are found in every season as they are grown almost all the year round, like most other tropical fruits; for in the tropics there is little seasonal variation, or at least there is no definite period during which plant life is at rest.

It appears that banana trees (*Musa paradisiaca*, *M. sapientum* and *M. nana*, to mention only the principal species of the Musaceae) originated in South Asia, more specifically in India, China and perhaps the Malay archipelago, where they were grown even in the very remote past. Their fruit formed the main source of food for many people. The Spaniards imported them to tropical America, and the Portuguese to Brazil. Later the French, who became major banana consumers, took the plants to their African colonies, to Guinea, Cameroon, the Ivory Coast and even the island of Guadeloupe. It is certainly one of the most widely grown tropical plants; indeed, there is virtually no tropical country in which it is not to be found.

The banana tree is a herbaceous – and not a

woody – plant, and is perhaps the largest of its type. It is fairly easy to grow and doubtless very profitable. Bananas may be eaten raw, cooked, or dried, and they are also used to produce a flour with a high starch and sugar content.

The main banana-producing countries are, in order of output: Senegal, the Canary Isles, Colombia, Guadeloupe, Martinique, the Somali Republic, Guatemala, the Central American States, Cameroon and Angola.

Like the banana tree, the sugar cane and the pineapple are monocotyledons. The sugar cane is a large graminaceous plant whose scientific name is *Saccharum officinarum*; it was first produced on an industrial scale early in the nineteenth century. Although sugar (sucrose) has always been obtained in tropical regions from sugar cane and other plants such as *Sorghum saccharatum*, Gramineae, palms and in particular the *Arenga saccharifera*, sugar cane was and still is the chief source of this product, which is so vital to man. The world production figure for cane sugar exceeds that for beet sugar in a proportion of approximately 8 to 1. Apparently sugar cane first came from India, where the local people used to suck the syrup out of the long stalks. To extract the sugary substance, however, the tender green shoots are bruised and the juice squeezed out.

According to old Chinese documents, solid sugar was produced only in 300 or 600 BC. And it was not until Alexander the Great's expedition to India that the first rumours reached Europe about a 'strange cane that produces a type of honey without the need for bees'.

First imported into Peru (in about the sixth century), sugar cane was subsequently grown extensively in the area; Peruvian scholars were responsible for research to find the best way of extracting and refining sugar, in other words the method of producing the white sugar that is still known in Peru as 'kand' – the derivation of the noun 'candy' or the verb 'to candy', as in 'candied fruits'.

In the seventh century, the Arabs – then masters of Persia – cultivated sugar cane in that area and then brought it to Egypt in the eighth century, as well as to Palestine, North Africa and even Spain.

From that time on, the consumption and in consequence the sale of sugar became of great importance in the whole of Europe. It is of interest that when Napoleon I declared the continental blockade, he told French chemists and scientists to find another sugar-producing plant

that would make up for the absence of sugar cane. They focused their researches on the beet, whose sweetening properties had already been discovered by the German Margraff, in 1747.

The sugar cane is now cultivated all over the tropics, provided that the average temperature does not fall below 20°C, and that there is a water-logged soil rich with nutritional substances.

Among the major sugar-producing countries are India, Cuba, Egypt, Taiwan, the Philippines, the Hawaii Islands, Puerto Rico, Brazil, Java, Australia in general, the Dominican Republic, Peru, South Africa, Argentina, etc, although America has the greatest acreage devoted to this crop. A residual substance, molasses, is used to make rum, a highly alcoholic drink very popular in the New World.

The pineapple (cultivated varieties of *Ananas comosus*, Bromeliaceae) is of American origin. This strange 'compound fruit' (a syncarpous fruit) resembles a large fir cone – hence its English name, 'pine-apple', and its Spanish name, 'pinas'. It is cultivated and marketed either for consumption raw or for conversion into fruit juices and syrups.

It was a long time after the discovery of America before the pineapple was first introduced into Europe; the Portuguese took it to India in 1594. In the eighteenth century, a Dutchman grew the first greenhouse plant in Europe, which was taken to the royal gardens of Versailles as a botanical curiosity. Today the pineapple is extensively grown in the hot regions of America up to the Antilles, in Hawaii and the Philippines, and also in Malaysia.

Although the fruit-bearing plants are of considerable economic importance, there are others which are no less vital: those, for instance, that have given rise to industries which in some countries may be the economy's major support. An example of this is the rubber tree, *Hevea brasiliensis* (Euphorbiaceae) in Malaysia.

This is not the only plant from which rubber is obtained; there are many other species, some Euphorbiaceae, some not, that supply this valuable substance. They are to be found in all tropical areas of the Old and the New World. However, *Hevea* is undoubtedly the most important and also the most profitable.

Rubbery substances were known and used by certain Asian tribes and peoples of tropical America in the remote past, although they were prepared in very primitive ways. History recounts that rubber was first made known in Europe by the historian, Pietro D'Anghiera of Arona in the early sixteenth century. As Campese writes, D'Anghiera was a councillor at the Spanish court during the reign of Ferdinand of Aragon. In his 'Decades of Things Oceanic and of the New World', he stated that the Mexicans 'had the habit of playing with balls made of an elastic material that they had obtained from a large forest tree'.

At the beginning of the eighteenth century, Charles de la Condamine brought samples of this 'elastic material' from Ecuador to the Paris Academy of Science, calling it by its native name of 'caoutchouc'. It had been obtained from a tree known to the people of Ecuador as 'hévé', and years later the botanist Müller von Argau was to give it the generic name of *Hevea brasiliensis*. The Brazilians cultivated what they called 'seringueira' on an extensive scale, especially in the state of Pará beside the Amazon, where indeed the plant grows naturally in large forests.

The rubber tree has long been cultivated in many other tropical regions of the Old World, as well as on the lower slopes of mountains in Cameroon, Ceylon, the Malay peninsula, Java and other Indonesian islands.

Hevea is often grown together with coffee and kapok (*Eriodendron anfractuosum*). It appears that the rubber plant produces more latex when combined with the other crops, especially coffee.

The other rubber plants cultivated are the Pará rubber tree or *Hevea discolor, H. guayanensis, H. confusa* and *H. benthamiana*, the first two from the Rio Negro and French Guiana, and the last two from Guyana and Venezuela; the rubber trees from the Brazilian state of Ceará, *Manihot glaziovii* and *M. violacea*, (Euphorbia-

ceae), the trees from the state of Pernambuco, *Hancornia speciosa* and *H. lundii* (Apocynaceae); other trees that are commonly found in Colombia such as *Sapium tolimense* and *S. thomsonii*, (Euphorbiaceae); the *Castilloa elastica* (Moraceae) from Central America which produces 'black rubber', and finally the African rubber plants, *Landolphia owariensis*, *L. klainei*, *L. droogmansiana*, *L. heudelotii*, *Clitandra arnoldiana*, *C. kilimandjarica* and *Funtumia elastica*, all of which belong to the Apocynaceae family, as well as the *Ficus vogelii* of West Africa, and the related species, *F. elastica* from parts of Asia.

By now it must be evident, through the descriptions and the illustrations in this book, that the largely unseen areas of our planet support an astonishing variety of plant life. Nowhere is too wet, too dry, too cold, too warm, too high or too low to provide a habitat for some plant species. And in addition to the strange and extreme environments, there are great areas of the world where vegetation flourishes in near-ideal conditions; yet few of us who spend our lives in temperate climates ever see more than a few isolated examples.

The importance of foods, drugs, cash crops and timber is such that the economy of whole countries has come to rely on the products of the land; families, towns and governments being dependent on the plant harvest and the world price for their goods. At the same time, plants that have no immediate economic value but which are an integral part of their ecosystems – and a pleasure to the eye – must increasingly be regarded as of the highest value; jewels in the devalued currency of our world environment.

In this account of some unusual and specialized plants, therefore, the object has been to familiarize the reader with the enormous wealth of our vegetation – its types, habitats, structure, classification and significance for man – and this diversity alone is enough to emphasize what a contribution the plant kingdom makes to our lives, how essential is its understanding, and above all how pressing is the need for sufficient attention to be paid to conservation.

For there are large areas of land – in the Congo, Amazonia and Brazil, to mention only three – where land is potentially suitable for cultivation by man, and yet where careful selection is paramount if the maximum number of species is to be preserved. While we admire the beauty and range of plant species, let us remember the future.

Bibliography

Bruggeman, L., and Campbell, W. M., *Tropical Plants and their Cultivation*, Thames & Hudson, London 1957.

Corner, E. J. H., *The Life of Plants*, Weidenfeld & Nicolson, London 1964.

De Wit, H. C. D., *Plants of the World: Higher Plants*, Vol. I (1966) and Vol. II (1967) London.

Eyre, S. R., *Vegetation and Soils: A World Picture*, Edward Arnold, London 1963.

Eyre, S. R., and Jones, G. R. J., *Geography as Human Ecology*, Edward Arnold, London 1966.

Galston, A. W., *The Life of the Green Plant*, Prentice-Hall, New York 1961.

Hill, A. F., *Economic Botany*, New York 1952.

Peattie, D. C., *Flowering Earth*, Phoenix House, London 1948.